more than the rare things

Rethinking biodiversity for human survival

Mark Dangerfield

Mindful Sceptics

Published by Mindful Sceptics

Springwood, NSW 2777, Australia

mindfulsceptics.info

AI Use Statement

This book was conceived, written and edited by a human who occasionally used AI tools to assist with grammar, editing, summaries and focused searches.

Contents

Preface

In the mid-1980s, I had a transformative and privileged experience as a postdoctoral fellow at the University of Zimbabwe. This foray into Africa, my first time living outside England, thrust me into a world of vibrant energy and hope. The air crackled with the electricity of a newly independent nation, its people buoyed by optimism before the looming economic challenges would cast their shadow.

Early in this African odyssey, I ventured into the heart of the Zambezi Valley, into what Westerners would call the real Africa. As the sun dipped below the mopane trees, painting the sky in hues of amber and crimson, I found myself standing on sandy soil, mere meters from a female elephant. The proximity was breathtaking, allowing me to hear the rhythmic grinding of her massive molars as she processed the dry grass.

Despite my intrusion into her world, she remained unperturbed, focused on her eternal quest for sustenance—majestic yet serene, powerful yet peaceful. Her wrinkled hide, a topography of time and wisdom, moved with surprising grace as she shifted her enormous frame. The gentle rumbles of her communication with the distant herd vibrated through the air, a primal language that seemed to resonate with the earth beneath my feet.

Watching this magnificent creature feed, I was given a masterclass in evolutionary adaptation. Her trunk, a marvel of nature's engineering, deftly sought out the most succulent tufts of grass, discriminating with a precision that belied its size. The elephant's role as a vegetation hoover and habitat modifier was obvious, her every action shaping the landscape as surely as millennia of wind and rain. In those precious moments, the pungent aroma of the savanna—a heady mixture of sun-baked earth, crushed vegetation, and the musky scent of wildlife—enveloped me. The majesty of the African wilderness, distilled into this singular encounter, was overwhelming. By some alchemy of the moment, I felt myself transported across time, briefly inhabiting the skin of our ancient ancestors. I stood there, metaphorically clothed only in a loincloth, a spear planted vertically in the sand beside me, connected to the timeless presence of the natural world.

This encounter was more than a mere wildlife sighting. It was a profound communion with one of nature's true marvels—the elephant, in her quiet dignity, embodied millions of years of evolution, a living link to the Earth's ancient past. As I stood transfixed, awash in the golden light of the setting sun, I was acutely aware of my privilege in witnessing this rare and beautiful moment, a fleeting yet eternal connection to the wild heart of Africa.

My time in Africa and my career as an ecologist gave me every opportunity to advocate for rare and endangered species. I could have become a campaigner, one of the saviours of these endangered souls. But I have not. Something felt wrong about saving the koala, the elephant, or whatever species was currently in the frame; something was missing. In *Missing Something*, published in 2013, I wrote about our vexatious love affair with the environment as I saw it from the disparate trenches of academia, business and a long practical experience of the environment. The book is worth a read, even if I spoil the punchline for you now. What I thought was missing was awareness. I imagined that all we needed to do was wake up, and all would be well or at least fixable.

A decade later, most things are worse. Every hour of every day, 8,000 more human beings join the global community of 8 billion, up from 7 billion when I wrote the first book. As our numbers and standards of living rise, we squeeze ever more resources from nature. Rather than this being success, we are told that we stand on the edge of an environmental cliff about to fall onto the rocks of global warming, pollution and starvation below. The cliff is steeper, the sea below is boiling, and we are much closer to a crumbling edge. And we still do not know if we will plunge or not.

Awareness was a nice, polite solution to the cliff conundrum, and, of course, it was naive. Humanity needs enlightenment, but that will not be enough; it needs more. We remain complacent, more so than a decade ago. Dissonance and the everyday mean we don't see the cliff, let alone the risk of falling from it. Instead, the rare creatures Sir David tells us are endangered are the focus of attention, especially saving them, as though if that were possible and achieved, all else could be forgotten. It can't, and it shouldn't. Even if all the rare things were 'saved ', humanity might not be.

Saving the koala, the whale or the elephant may fuel the dissonance and protect us from guilt, but it is not the answer to the future of humanity, no matter how desirable such an outcome is. And yet, it is more than just saving; I contend that passion for the rare is part of the problem, a big part. No, the future of humanity and a livable environment is much more than the rare things. And yet, it is not obvious why. This book explores our fixation on the rare and why we need so much more. It is a reminder that rare things are inevitable and that extinction is a key element of the evolutionary process, even if human activities have dramatically accelerated the rate of species loss. This might be a difficult truth, but the current fixation with the rare will not feed everyone or maintain the natural resources needed and wanted by 8 billion humans, all looking for betterment. Human survival needs a paradigm shift in our conservation and environmental stewardship. This book explains the why and some of the how.

No author writes alone. I fed off the insights, wisdom and encouragement unconditionally offered by Chris Scott, my co-founder at Mindful Sceptics, and Charlie Wilson, who built AfterBefore with me. Thank you, gentlemen. Several mentors, teachers and friends have contributed to the ideas in this book long before I wrote them down. My gratitude and respect to Mark Hassall, John Ingram, Mike Swift, Vic Clarke, Bruce Riddoch, Jem Perkins, Woody Cotterill, Diana Wall, Brian Springett, Andy Beattie, Shona Bates, Chris Allen, Bryce Wilde, Julie Busuttil, Kristian Holz, Kirsten Prior, Anthony Pik, Duncan Tyler, Gregor Schuurman, Samson Kaunda, Ian Oliver, Phil Saville, Kim Crawford, John Szymanski, and especially the late Steven Telford. Naturally, all the opinions in this book are my own. And without the love and support of my wife, Milena Muscatello, none of these thoughts would have made it out of my head.

More than the rare things, it is meant to be entertaining and informative, but it is also a little less nice because these are challenging times. I hope you still read it and gain from a little discomfort.

Chapter One

Birdwatchers

Jason flew over the dry stone wall and ran like crazy across the field, scattering the sheep. Mike, Jason's best mate, was less agile over the wall but only a couple of strides behind. I stooped to pick up the flat stone that Mike had dislodged. There was no cement or mortar in these ancient barricades and I guessed it must take a long time and considerable skill to build them.

We were near the end of a memorable week in Devon, the sleepy and always very rural-looking corner of England, the home of scrumpy and cream teas. As youngsters only recently let off the parental leash, we had woken the worse for too many pints of the unique rocket fuel the locals make out of apples. Every morning, I vowed never to touch the stuff again, only to lose my resolve as I stood at the bar in the evening. Drinking was a competition between males to impress and imbibe Dutch courage. Only there had to be more to it, as the girls always ignored our approaches. We were all much older before realising they had been paying close attention the whole time. We did not indulge in the double-clotted cream, thick enough to mould into a small statue or fill in the gaps in the dry stone walls. Such physiological challenges were left to the tourists and their bigger wallets.

We were freshmen on a field trip in ecology, the subject that makes a science out of how animals and plants make their living in the world. Back then, few students studied ecology, at least not compared to those learning how humans make a living, but you did get to go to Devon. Today, there are still too few devotees, and they have let the field trip slide.

On our first day in Devon, we trekked to Wistmans Wood, a delightful place that hugs the side of a hill on Dartmoor. It was one of those rare spring days in England when the sun shone, and it got warm enough for the girls to reveal their delectable shoulders. Tradition has it that Wistmans Wood is haunted by a pack of Yeth hounds that sortie from within the moss-covered boulders and twisted trees to hunt for human souls, howling in unison as they follow their master across the moorland. We were not told of hounds, wraiths or any other ancient evil. Our instructor, an academic sporting the bushy beard compulsory for all real ecologists, knew this particular patch of countryside very well. It was the subject of his doctoral thesis. He told us all about the unique interactions between oak, moss, and boulder that would always result in a species' poor combination that bucked the convention of stability from diversity—no mention of the devil.

After the blissful sunshine and intellectual repartee of Dartmoor, we got wet and very cold. Such is the nature of the weather on the windswept isles and the somewhat masochistic desires of limnologists. Our next instructor anticipated our discomfort, insisting that all first-year students measure biological oxygen demand in a large expanse of fresh water. BOD, as it was affectionately known, is a measure of the productivity of a lake, and back in those primitive days, when lecturers delighted in getting students to measure things the hard way, BOD required the immersion of two bottles of lake water several feet below the lake surface. One bottle was covered with duct tape to make it opaque, and the other was clear. Placing and retrieving the bottles required navigating the lake in an unstable inflatable boat, ensuring everyone got wet and chilled to the marrow. Measuring the oxygen levels in the bottles via titration back in the laboratory gave you the rate of photosynthesis. These days, you get the same BOD results in a couple of minutes with a probe. This is fortunate because occupational health and safety regulations prevent boating by infants.

The day that Jason and Mike disappeared over the wall, we were on a walk to a beach. It was blustery, about halfway between the sunshine of Wistman Wood and the BOD afternoon. The gum boots we were instructed to wear explained Mike's clumsiness over the wall. I felt good, a passion for this ecology thing stirring somewhere above my loins, yet the rocky shore option did not appeal much. I guessed that we would be mapping the already famous zonation of shellfish between the tide lines. The example appeared in all the textbooks to prove the concept of a niche, defined as an organism's living and doing space, a pillar of ecological thinking. On the pebbly shore, the tide's ebb and flow creates a balance of immersion and exposure, shifting towards the latter as you move closer to the high tide mark. As the English Channel is part of the shallow and constrained North Sea, its wide tidal range extends the exposure zones and makes the horizontal pattern easier to see.

We were told, as though it had been written in the gospels, that the zones were maintained because of a process called competitive exclusion. However, you can also get these patterns without an aggressive process. Today's ecologists call this method 'neutral theory'. I suppose that even the marble pillars of the Greek temples crumbled, but there was no doubting the pattern. It was clear, and the given explanation was considered a no-brainer. At the time, I put my indifference to rocky shore capers down to a preference for dry habitats or places where gum boots repelled mud and not water. Later, I realised it was a subconscious concern with acts of blind faith that a given pattern came from a given process.

Once over the wall, Jason and Mike maintained a healthy gallop across the grazed pasture. Jason had ignored the gumboot imperative and was first to the far side of the field. Curious at the bizarre behaviour of our comrades, we had all stopped to see what would happen. Jason halted at the opposite wall and retrieved a well-worn book from his backpack. He thumbed through the pages and consulted it between bobbing his head over the topmost stone. Mike soon joined him, and they exchanged excited gestures, which resulted in Mike also bobbing. After a minute or two, they both scaled the second wall and disappeared from view. A graduate student tutor named Elias grudgingly accepted the task of retrieving the runaways while the rest of the party continued to the beach.

The English countryside is designed to be on picture postcards. The colour, texture and order of agriculture meld into an intimacy you cannot find where food is grown on big, monotonous fields. If only the postcard makers could bottle the smell of earthy freshness with hints of damp soil and the delicate blend of floral sweetness intertwined with the crispness of dew-kissed grass. The lane down to the beach was ancient, like the wood and felt permanent. Even as a bawdy youngster, I felt a sense of place. The tarmac took something away, but the lane had heard the horses, felt the rattle of wooden cartwheels, and smelt the manure as, year after year, crates of shellfish were transported to the village from the beach where generations of rugged fishermen in oilskins hauled rowing boats onto the shingle without noticing any of the ecological zones.

Beyond the dense hawthorn of the hedge, the fields also had permanency, with *Festuca, Poa* and *Astragalus* grasses all meshed together cheek by jowl, even with, or perhaps because of the attention of sheep with ancestry as long as the people. Abundant broad-leaved herbs kept a low profile as their nitrogen-rich leaves fell to provide a vital source of high-quality food for the detritivores, who kept the system productive in their turn. Since that time in Devon, I have walked or flown over many agricultural landscapes on four continents and the occasional island. Nowhere has felt as dependable or as secure as that landscape in Devon. English fields have been fields for centuries, the wise old men of sheep production.

We ambled down the lane to reach the beach and gathered in a huddle to listen to the lecturer explain what was required of us. I had guessed right; it was zone mapping day. Rather than provide precise instructions, we at least had to figure out how to generate the result. We teamed up into groups of three and, rather like certain European tourists, claimed a stretch of beach as our own.

It is amazing how like attracts like in the selection of team members. Some trios formed instantly and cracked straight into it, all gung-ho with tape measures and quadrats flying around with the freedom to have a life of their own. Then, there were trios of meticulous planners. These folk negotiated team membership, discussed what to do at length and wrote down their plans in excruciating detail. It wouldn't surprise me if one of the meticulous groups is still recording mollusc data to this day. I later learned that the best science requires balancing the gung-ho and the meticulous. Too much of the former and you get a sloppy, inconsistent and ultimately unsatisfactory result; too much of the latter and everyone goes home before a result appears. Because good learning has many layers, that day, I realised that even the simplest or most familiar exercises can expose more than one of them.

It was lunchtime before Jason and Mike, escorted by a still disgruntled Elias, reappeared, beaming and showing signs of close encounters with the muddier parts of Devon. The boys were 'happy as Larry' to use a very English expression.

"Awesome," Jason barged into our lunch circle in the way only very excited people can, "we saw one, and then a pair, way too cool." When no one replied, he filled the gap. "If they were breeding, that's the first record for twenty years, Bren will be well pissed."

Brendan was Jason's usual partner in crime. Still, given they had surnames at opposite ends of the alphabet, today Bren was on the dunes with another student group, wondering about the ecological function of marram grass.

My trio were lunching with one of the gung-ho groups who had, as expected, finished first, and Jason's intrusion had halted what was, for freshmen, a rather intellectual discussion. Megan, an especially attractive gung-ho who had not been impressed with my expression of Dutch courage the night before, had found a weird invertebrate in a rock pool. Although her trio had not left frantic mode, she had recognised that the critter was unusual and was smart enough to put it in a specimen jar for a closer look.

"It's a Nudibranch, a sea slug," I said, knowing full well that I had let myself in for serious nerd demerits, but I added a smile that I hoped might balance it out. I had passed many hours the previous semester in the company of weird spineless creatures and the unique smell of formalin; such was the traditional invertebrate zoology class. One of the pickled specimens we were forced to draw was a Nudibranch, a family of opisthobranchs within the phylum Mollusca. Why we had to draw one, I couldn't fathom. The last art lesson I took was in fourth grade, but I was told it would sharpen my observational skills and, as it turned out, gain nerd demerits.

Megan's live one was glorious in comparison to the pickled versions in the glass cabinets of our zoology lab. It looked like a tiny pink hair brush with a semi-circle of thick plastic combs, only in this animal, they waved about and had white tips. Megan passed the jar to Elias, who had sat down next to her. His reaction was almost as extreme as the earlier hurdling of the wall. He sat bolt upright from the slouch he had brought with him from his chaperoning chore and exclaimed, "Extraordinary! Amazing."

"What?" Megan asked.

"This is only the second one I've seen in three years. Amazing. *Flabellina pedata.*"

"A nudibranch?" I asked.

"Yes indeed." Elias stared at the jar.

"Is it rare?" Megan asked as though she already knew the answer, which explained everything.

"Not really. Locally abundant is the best description. I just haven't seen one here. Where was it?."

"In that rock pool over there." Megan pointed in the vague direction of the gung-ho patch of beach.

"Truly amazing..."

The interaction seemed less exceptional when Elias revealed that he had spent three years sampling invertebrates, especially nudibranchs, along the rocky shores of southern England, data that he would turn into a doctoral thesis. Megan's nudibranch was rare in his dataset, but when you found one, there were often many.

This news added another layer to our earlier intellectual banter. We discussed why humans always notice the unusual. It was as though everyone had an eye for it. Phil, one of my trio, suggested it was because our food was best found by having a search image for the unusual. Fruit, tubers, and rodent burrows were more likely to be found if we searched for something different. Phil had recently completed an animal behaviour essay and was therefore bold enough to suggest that this search image idea may have contributed to the success of humans as a species, especially in our hunter-gatherer days. We all nodded and agreed he had a point, an acknowledgement that allowed him to smile at Megan. How Elias reacted to the rarity of the nudibranch seemed to lift the idea to the lofty status of hypothesis, yet tragically untestable. A learning feeling was palpable.

Jason couldn't contain himself any longer. "Don't you guys want to know what we found?" Answering his own question after the briefest pause, "A hoopoe, you dumbats, maybe a breeding pair. It's a lifer for half the birders I know." I must confess that I found it hard not to catch Jason's enthusiasm. He was rapped at his good fortune and proud of his skill in recognising the bird that flashed across his peripheral vision as something unusual. I also couldn't help but be a little impressed.

In later years, I came to think that bird watchers lived for lifers, this bird species they see for the first time, perhaps the only time, often after many years of searching. I did wonder what it was called if you saw a lifer a second time. Some years later, I was reminded of Jason and Mike's antics almost daily. As a golf tragic living in southern Africa and on the course as often as possible, there was a hoopoe on every fairway. The bird is unmistakable once you have seen it: about the size of a dove with rufous coloured feathers on the head and body, pied wings, black tail feathers and a thin beak, appropriately curved to probe the turf for insects as it walks along. If you come too close to one, it flashes a conspicuous crest of feathers in disgust. The bird I saw on the fairways taxonomists classify as *Upupa epops*, the same species Jason and Mike chased in Devon. What made it so exciting for them is that it is a warm climate species that does not breed in England. The individuals they had seen were vagrants from northern France, most likely blown over the channel in a storm. Today, with the globe warming up, Devon might be attractive enough for them to settle down.

In Africa, I was surprised to see hoopoes in such numbers. A little of Jason's excitement came to me the first time I saw them. After a few fairways, I enjoyed seeing them but didn't feel the need to jump around or scale barricades to get a closer look. Familiarity did lead to complacency. I knew hoopoes were not rare in southern Africa; you just had to look for a manicured lawn.

I lived in Africa for nearly a decade and would return to England periodically to visit family and remember the cold, but I never felt like returning for good. A couple of days in the London traffic and I was over the congestion and yearning for hoopoe habitat. On one visit to my uncle in north London, where there were few wildlife opportunities but plenty of enthusiasm, we went to walk in an area where ponds and canals had been rehabilitated and landscaped into a user-friendly wetland reserve. We strolled along paved walkways and boardwalks and ducked into wooden bird hides. All manner of people, rugged up against the winter chill, were out and about, some pausing long enough to read the excellent interpretive signs. There were bird watchers aplenty, almost as common as the starlings. Most came to see the water birds: herons, grebes, various types of ducks and especially a pair of bitterns. The human watchers of these creatures were like mature versions of Jason and Mike, the enthusiasm still innate but somehow tempered by the behavioural norms of older age.

A week earlier, I had been in the Okavango Delta in northern Botswana, a RAMSAR-listed wetland that boasts more than two hundred breeding bird species. I was working on a termite project, setting out, in neat grids, blocks of wood to find out where these soil critters go to find food. Even with my head down, I had managed to see feathered rarities such as African skimmers, wattled cranes, Pel's fishing owl, and Heuglin's Robin, among others, along with half a dozen species each of egrets, herons, geese and ducks. In the sky were Bateleurs, kites, snake eagles and various vultures that took some skill to tell apart. I didn't see a bittern, but it was considered a resident of unknown status. I couldn't bring myself to crow as my uncle, like the other green-clad folk with spotting scopes and binoculars, was as enthusiastic over the bittern in London as Jason and Mike had been over their vagrant hoopoe. It was a bit like the pride of a baby boomer when his son is old enough to appreciate that his dad actually went to a Rolling Stones concert in the 1960s. Mine didn't.

"A pair of Great Crested Grebes nested in the reeds last summer", my uncle said as he pointed towards a nondescript clump of vegetation on an island that resembled a building site in the rain. I suppressed an image of towering papyrus stems, a deep organic green in the orange light of an African sunset.

"How many breeding pairs are there?" I asked.

"I have seen two, and we also get swans and grey herons." I was not sure if he meant that they were here or that they were breeding.

"How often do you get out here?" I said.

"Oh, not as often as I would like, maybe two or three times a year. They have done a lot of work, though. It used to be such a mess."

"The hides look good"

"Yes, they put most of them up last year too. It can get a bit crowded—see, over there, a little grebe."

I struggled to make conversation. It felt like my uncle was trying to teach me something I had learned in kindergarten. He wasn't trying to impress or be something of an equal. He knew about some wildlife on his turf, and he imagined I didn't. As far as he was concerned, I knew about Africa. It is easy to forget that someone might have similar knowledge.

In the car, on the way back to his house, I thanked him for showing me around his wetland and not worrying that we didn't see the bitterns. He smiled. A grin that got wider at the point I called it his wetland, as even armchair enthusiasts like to feel a sense of ownership of wild places. And if you live in the suburbs of a city of eight million people, a semi-restored patch of wetland the size of a few football fields is wild enough. Even though my uncle tolerated my indifference to his favourite pastime, he could not resist asking me about the rare bird species I had seen in Africa. He had guessed correctly that I had not spent all my African days in my office. I told him the hoopoe story of Jason and Mike and how I saw hoopoes almost every day. It was a lifer for my uncle, too. He wriggled with excitement.

I thought it better not to mention the skimmers, cranes, fishing owls, ospreys and fish eagles. It didn't seem right.

The Royal Society for the Protection of Birds in the UK has over one million paid-up members, making it the largest wildlife conservation charity in Europe. This is a more significant number of members than most political parties could boast in their heyday. Perhaps people are more interested in the whereabouts and habits of Reed Warblers than they are about who should run the country. I agree that is a stretch, but a lot of people like birds.

Roughly 620 bird species are recorded in the British Isles—the actual number varies from year to year—and seeing them all in the wild requires time, travel and no small amount of luck. There might be a dozen species in your backyard, perhaps another 20 or more in the park gardens and hedgerows around the district. A short drive to the local reserve and the elusive Reed Warbler might be spotted. That is if you live within a short drive of half a dozen places where the habitat is still in good shape. Should you want to add *Tetrao urogallus*, the Capercaillie, to your list, then a trip to native pinewoods of the eastern highlands of Scotland is required to try and find one of 500 breeding pairs. Most birders fork out hefty sums on travel. RSPB members pay the equivalent of USD75 in membership fees each year and hundreds more on bird-watching paraphernalia. The U.S. Fish and Wildlife Service estimated that bird watchers in America contributed $36 billion to the economy every year, mainly on the dozens of specialist tour operators who organise itineraries to far-flung places to see rare and exotic bird species.

Birdwatching is a great pastime with all the elements of a rewarding hobby. First, it requires getting out of the house and into the wilds. This adds the dual benefit of exercise and communion with nature rather than Netflix. It also allows you to shop online for a muted weatherproof jacket, some hiking boots, and light but powerful binoculars. As your interest grows, this could extend to monopods and telescopes, perhaps a camera with a telescopic lens. Every walk or visit to a hide has a sense of expectation. What will be there? What will be seen or only heard? Details of song and calls and plumage must be memorised, which is always a good way to keep the mind active, and the list can never be full because the juveniles, hybrids, and elusive vagrants will always keep you on your toes. All this leads to the animals themselves.

Not only are birds pretty, but they fly. This is extraordinary, and each species has its way of defying gravity, from soaring raptors to flitting flycatchers and hovering hummingbirds. Some other lazy doves and pigeons prefer not to unless pressed. Each species has unique behaviours to forage, preen, avoid danger, and socialize or not. Hours and hours of observation are needed to understand what each species does on its daily round. Birds build a symphony of sounds, ranging from lilting and rhythmic chirps to hauntingly beautiful melodies and loud screeches, each species contributing a unique call. The songs convey territory, courtship, and communication messages, filling the air with a vibrant chorus. Then, there are the interactions between the species. Who bullies who, and who tells the bullies to take a hike?

At the moment, I live in a suburb in the Blue Mountains west of Sydney, Australia. My garden has bushland and the usual lawns and a jumble of garden plants near the house. What exactly is the pecking order in my backyard? The lowest rank among regular visitors goes to the Brown Wood Pigeons; almost everything chases them away. Next comes the King Parrots, who get their name from plumage rather than respect. Then, there are the Crimson Rosellas, although they are not really regulars, and the Sulphur-crested Cockatoos, who push every other species away with sheer bulk. The lords of the bird table are the Rainbow Lorikeets, a pair that has a nest in the hollow of a spotted gum in the neighbour's yard. They are not only more colourful than seems possible, but they are fearless and feed at their leisure by sheer force of personality. Away from the table, the Noisy Miners rule the skies. A cooperative breeding species that gang up on everything that dares come anywhere near them. This keeps the Spinebills, Fairy-wrens and Thornbills as occasional visitors to keep the Kookaburras and the Butcher Bird company.

Then, in the summer of 2019, a huge wildfire for weeks burnt through more than 500,000 hectares of forest to the north of our suburb. For a few days, we had a vagrant pair of Lyrebirds in the gully below our garden, braving the neighbour's dogs and deciding if they could tolerate substandard temporary accommodation. They couldn't and moved on, but not before giving us the pleasure of their crisp, varied calls for several days. Other woodland species stayed on for longer, including the extraordinary plumage in sexual dimorphism of the Bowerbird; the male is metallic black purple, and the female is mottled green. The ever-frisky Honey Eaters took up residence, and there was even a Friarbird for a day or so.

Despite all these sightings and the morning visits of the King Parrots and Lorikeets to our deck that overlooks the eucalyptus trees, I am not a bird watcher. I learned their names because I'm a scientist trying to observe nature. That is the prime skill of any ecologist. I must have seen several hundred species in my sixty-plus years. The decade in Africa notched up a few, as did the mountains of Papua New Guinea, the Indonesian islands, and other exotic and more mundane places I visited over the years. When arriving at a new place, the main plants and birds appear in the neighbourhood to jump onto our lists. It also works to feed one of our core skills, the ability that did much to secure evolutionary and numerical success. The observational power is to recognise the unusual and then to fix the search image for aspects of that unusual, whether it be good food, dangers or one of the unusual species of bird not yet ticked off your lifetime list.

This kind of list would annoy the true ornithologist with a serious inventory on his iPad. The birds he has seen, where, when and something about how many? A pair of Sooty Oystercatchers on the New South Wales coast near Hawks Nest was one of mine and a Shoebill on the reeds deep into the Okavango Delta. Yep, me too. They make the list of this lifetime. How serious he is will depend on where he chooses to go. I have no South American species on my list unless you count those toucans in the zoo, and of course, you cannot. Any true list must be from specimens seen or heard in the wild. How much rarity I can observe depends on where I am prepared to go.

Jason and Mike scored a lifer, but Hoopoes are not rare; they occur across Africa, Asia and Europe. Three living and one extinct species are recognised, though for many years, all of the extant species were lumped into a single species, *Upupa epops*—a species only rare if you don't know where to look, or as this last piece of information hints, you may not be exactly sure what it was that required a leap over a dry stone wall and chase across a meadow. Taxonomists now believed there are three species of Hoopoe, not one.

Rarity then has any number of sources. There are few individuals in a given place, such as the Devon Countryside, but many elsewhere—rare at the edges of a range but common enough within it. Such species are easily enough to add to the list on your master bird watcher's spreadsheet.

Then there are rare because of a change to an especially nuanced taxonomy. In the case of the hoopoe, what was one is now three, increasing the rarity of each.

If you want to add a bird of paradise to your list, that would mean an extensive rainforest sortie because these species only occur in one spot on the planet. The same can be said for a host of other species, including Darwin's finches, which are famous for their contribution to the theory of natural selection, which is the best scientific explanation for so much rarity. More on this later.

Then, it is rare because natural selection produces a specialist species that can only persist in certain types of habitat, and if that habitat is uncommon, so is the species. Birds are not the best examples of this because they fly, allowing them to move between habitat patches.

These reasons for rare are a human abstraction, our taxonomic systems and how we choose to apply them, and finding specimens at the edge of a species' normal range. In the case of the hoopoe, blown off course in a storm and finding sanctuary on the nearest patch of terra firma. Rare from a narrow distribution, not so rare if you can find yourself within that narrow range, but rare according to the bird watcher called Eric, who lives in Accrington Stanley.

Another reason for rarity is a fundamental principle of energetics. In 1980, Paul Colinvaux wrote a book about it, and the title gives away the premise: *Why Big Fierce Animals are Rare*. The short answer to the spoiler of the book is that the laws of physics and chemistry mean that energy is lost at each trophic level in the food chain. The further away from plants an organism is in the eat-and-be-eaten sequence, the less energy available, so the fewer individuals the available energy can support. Rarity can come about because of diet. Huge flocks of Starlings, not so many Hen Harriers. Queleas by the million, not so many Yellow-billed Kites. Physiology also explains why there might be plenty of big birds, Ostrich, Emu, Cassowary, and even some that can fly like Bustards and Condors, but they will never occur in flocks of thousands. The food supply cannot support those numbers or the bulk.

All these sources of rarity are a tremendous boon to our man Eric from Accrington. He is always just a short drive and a bit of luck away from sighting a rare bird species. Thanks to hundreds of years of bird watching and record keeping by amateurs and professional ornithologists, he knows where to add to his list of lifers. These species are so elusive that Eric expects to see them only once in his lifetime. We shouldn't forget that last point. Some species are not rare but appear to be so because they are hard-to-see species, like Reed Warblers, which prefer dense reed beds, or one of the 40 species of Tucans from Mexico to Paraguay, which live high in the canopy of rainforests. Then there are cryptic species, which are so because they look very similar to other species. The LBJs, little brown jobs, often fall into this category. They are rare because of the difficulty of identification.

The smallest bird in the world is the Bee Hummingbird, *Mellisuga helena*, a native of Cuba. It is truly tiny, weighing just 1.9 g, but still visible to the naked eye. Making a complete catalogue is not a problem that requires microscopes or expensive technical equipment. Birds fly or scuttle around and are visible to us, which is a big reason why birdwatching is popular.

What all this means for Eric is that he has a passion for life. There are 632 breeding species in the British Isles; it will take some dedication to see all of them in the wild. He can extend his search to Europe and add another 268 species. A trip to Australia with its 828 species will add at least 380 more because they occur nowhere else, although some of the 828 he will have seen back in and around Accrington. Again, more on that later. Should he wish to keep going, we think the global list is around 9,000. The best published estimates of the number of described species are 9,000 or 9,750 or 9,875 or 9,917 or 9,946 or you could go with the IUCN round number of 9,990. Take your pick. This should be a surprise in the modern age of phone cameras, storage clouds, and 8 billion souls who have all seen birds.

Suppose we went to an international birdwatchers convention. The sort of gathering where enthusiasts share stories of adventures in Patagonia in search of the Yellow Cardinal or the endemic Band-tailed Earthcreeper. If we ask delegates to share their lifetime list of species seen, there is a fair chance that almost all the birds in the world would be seen by at least one delegate. We might even tick them all off with the handful or so of the most experienced campaigners.

Only there isn't a single definitive catalogue of all the bird species in the world. It's not a problem if there was an agreement on the species that go onto the list, but real life is never so simple. There are synonyms and taxonomic disagreements, duplicates and splitting of single species into many. The reality is that even for a group of animals with more adherence than the political party in power, that even a child can see with their naked eye, we don't know how many bird species there are on earth today. The indictment creates a problem for the delegate wishing to complete the global list in his or her lifetime. They don't know when they reach the end because there is no definitive list.

The Common Gull or Sea Mew (*Laris canis*) will be on the list of all birdwatchers from Europe and North America because they have seen this species many times. Not in the millions, but in small flocks and individuals almost every time they visited coastal habitat, hoping to view a rare wader or a vagrant Hoopoe. This species is classified as widespread even though it is a habitat specialist of coastal regions. Moving inland at certain times of the year into farmland, the Common Gull occurs over a large geographic area, from the Palaearctic across Europe and North America, that any committed bird watcher will have seen one. Experts estimate there are at least a million pairs across Europe.

However, the rusted-on birdwatcher would not be satisfied with ticking *Larus canis* off their list, for the taxonomists decided to split this species into four subspecies, two of which are considered distinct species by some authorities:

L. c. canus – Linnaeus, 1758 – common gull. nominate, found in Europe and western Asia. Small; mantle medium grey (palest subspecies); wingtips with extensive black; iris dark. Wingspan 110–125 cm (43–49 in); mass 290–480 g (10–17 oz).

L. c. heinei – Homeyer, 1853 – Russian common gull. Found in central northern Asia. Medium size; mantle dark grey (darkest subspecies); wingtips with extensive black; iris dark. Mass 315–550 g (11.1–19.4 oz).

L. c. kamtschatschensis – Bonaparte, 1857; syn. L. kamtschatschensis – Kamchatka gull. Found in north-eastern Asia. Large; mantle medium-dark grey; wingtips with extensive black; iris pale. Mass 394–586 g (13.9–20.7 oz).

L. c. brachyrhynchus – Richardson, 1831; syn. L. brachyrhynchus – mew gull or short-billed gull. Found in Alaska and western Canada. Small; mantle medium-dark grey; wingtips with little black and much white; iris pale. Wingspan 96–102 cm (38–40 in); mass 320–550 g (11–19 oz).

So to see all four subspecies, the intrepid delegate would have ticked off the common gull at Blackpool Beach, not far from Accrington, taking a quick trip to Leningrad for a few Baltic species and the Russian Common Gull, made sure that he included Alaska in his North American odyssey to tick off the Mew Gull and while there take a quick trip across the Bearing Sea to the Russian side again and find the Kamchatka Gull on the peninsula of the same name. As you can imagine, committed birdwatchers are great venturers and must also be excellent at geography.

The Indian Myna is a starling that has adapted wonderfully to urban environments. Originally an open woodland species able to eat just about anything, the IUCN considers it invasive because it is aggressive, territorial, and always hungry and will reduce the habitat quality for other bird species. The native range of the Indian Myna is central Asia, but it has been introduced into many countries. All an avid bird watcher has to do to see this one is to visit Canberra, the capital of Australia, where they survive at densities of 75 per square kilometre. It would be very hard not to see one. The range of the common myna is increasing at such a rapid rate that in 2000, the IUCN Species Survival Commission declared it one of the world's most invasive species and one of only three birds listed among "100 of the World's Worst Invasive Species" that pose a threat to biodiversity, agriculture and human interests.

What makes birds common is similar to what makes any other species common: an abundant food source with a reproductive system designed to exploit it, a flexible lifestyle, and adaptive characteristics that allow for a wide geographic range. A good scrambler or an aggressive contestant in the endless competition for resources among species. Common bird species can be opportunists, strong competitors, or adaptable species that can make a living off anything, perhaps a combination of all three.

At the birdwatchers convention, one of the delegates, Julie, a young woman from Wagga Wagga in southern New South Wales, decided to go about birdwatching differently than most. As a youngster, she was mesmerized by a flock of budgerigars when she visited the Menindie Lakes deep in the Australian Outback, tens of thousands of birds, far too many to count. This species is found wild throughout the dry parts of Australia, where it has survived harsh inland conditions for over five million years. Its success can be attributed to its nomadic lifestyle and its ability to breed while on the move. The size of the flock was what captured Julie's interest. There seemed to be nothing to eat in the parched landscape, yet the flock was thousands of individuals, all in their breeding plumage and best health. Intrigued by the idea of large flocks, Julie decided to see all the common species she could in a year of global travel rather than track down the rare types in the inaccessible and difficult-to-reach places with no guarantee that the elusive warbler would sing or be seen. She figured that by moving from place to place and covering as many habitat types as possible, she could get a healthy list of species seen. She mapped out an overland route to make the journey pleasant and decided to buy a car in each country, where that was cheaper than transporting one across the water. She saved and budgeted carefully in the year of travel and saw 4,322 bird species, just shy of half on the approximate global list. In a year, she had reached numbers that some enthusiasts do not achieve in a lifetime of trying.

At the reception before the conference dinner, Julie found herself talking to an American couple, Bob and Mary Lou from Thibodaux, a small town west of New Orleans, who did most of their trips together and could claim to be the two humans alive who had seen the Mississippi Sandhill Crane, *Grus canadensis pulla*, more times than anyone else and also knowing where to see Yellow-breasted Chats, *Icteria virens*, in winter, a species where almost all the population departs for warmer crimes in the south. Also in the huddle was Eric, our self-styled British ornithologist from Acrington who, in retirement, was a full-timer on the circuit. After an exchange of greetings, as happens among strangers at conventions, the conversation went something like this.

"So you're new to it. That's nice," said Mary-Lou.

"Well, always been fair dinkum about liking birds, but I only decided to make a big list two years ago," Julie said without realising where such an admission would end up.

"That is splendid," Eric said, itching to let slip that he had seen 547 of the 620 bird species recorded in the British Isles, including the recently arrived white-rumped swift.

"We were so lucky," Mary Lou went on as Bob became her nodding head. "We found a breeding pair of Mississippi sandhill cranes, probably the only ones still in Louisiana. That wasn't really the lucky part. We passed by a house on the outskirts of Thibodaux just a short walk from their swamp, and Bob just happened to see the for sale sign. We put an offer in that very evening, oh my, it's been so wonderful to see them every day."

Splendid, splendid," Eric said. "Did you know that the white-rumped Swift is now on the UK list? I saw one over a cornfield in Oxfordshire just last week. That puts me up to 547 for the UK. Not too shabby, hey."

"That's swell," Bob said like he meant it.

"I doubt you manage that many on your travels, young lady," said Eric.

"I think I'm up to 4,322 right now." Julie said, at which point Eric dropped his glass of beer on the carpet, and Bob said, "Gosh".

This fictitious anecdote tells us a lot—not just about rarity but also about how the human brain is wired. There are plenty of Erics and Mary-Lous among us who are either deluded about the importance of the tail of their list or the specific species that reside there. It also helps to notice that there are few Julies with the pragmatism to realize that if it's numbers you're after, there is a much simpler way than chasing the tail.

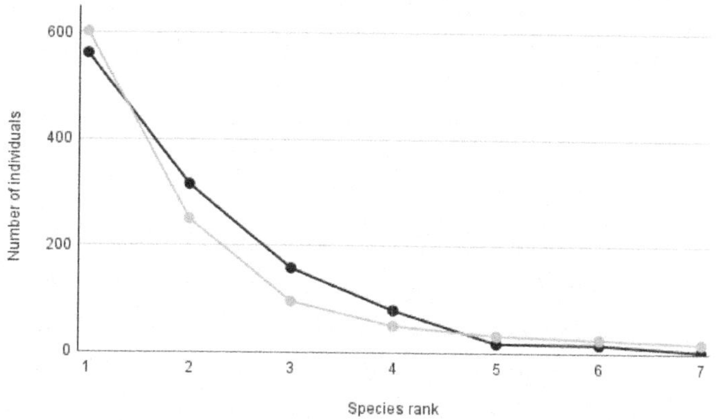

Rank-abundance plot for millipede species in southern African savanna habitats. Redrawn from figure in Dangerfield, J. M., & Telford, S. R. (1992). Species diversity of Julid millipedes: between habitat comparisons within the seasonal tropics. Pedobiologia (Jena), 36(6), 321-329.

First-year ecology students get this curve. They will have thrown quadrats in meadows and sampled grasses and herbs to generate such a curve from their data. While at it, they will also use the same dataset to plot this graph, which explains the first.

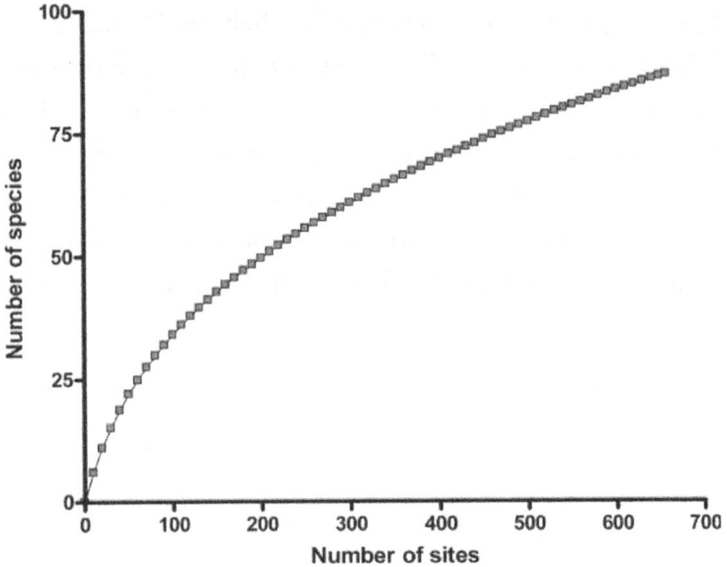

Species richness, in this case for millipedes in the south of Western Australia, increases with the number of sites sampled, sampling effort. Source: Moir, M. L., Brennan, K. E., & Harvey, M. S. (2009). Diversity, endemism and species turnover of millipedes within the south-western Australian global biodiversity hotspot. Journal of Biogeography, 36(10), 1958-1971.

It can be plant species in a grassland, fish species in a lake, tree species in a tropical forest or microbial species in the soil, and the same general pattern will appear every time these two curves are used. This is because, as Julie figured out, some species are rare and not easy to find until many samples of the habitat have been taken. However, half the sampling effort will deliver more than half the species in the habitat. Julie got to have a long list of species by applying this rule and moving on quickly to cover as many habitats as possible rather than waiting to put in a lot of effort for those rarely sampled species.

Erics are fixated on finding out how long the tale of this distribution will be. They forgot the common types many moons ago and constantly seek information to ensure they have all the possible species. Their focus is always on the tail of the distribution. Fixation with the size of the total number also makes Erics prone to splitting. Not their khaki pants but species. They are subspecies boys, always looking for finer divisions to add to their list. These folk might be the ones who have established dialects in the bird song, the passerine equivalent of Geordies and Scousers. We will come back to splitters and their lumping cousins later.

Mary-Lous don't even know the shape of the curve. They have found their rare darling that no one else knows about and stay with their species or, more likely, their specimens. Some become attached to specific animals like pets, give them names and would be happy if they could take them home. Not that you could get a crane into your SUV. Other Mary-Lous undertake the most intricate studies of their chosen rarity. They detail feeding, mating, migrations, nesting, and communication with every aspect of the bird's existence, often recorded and reported as detailed portraits of the bird's life history. These monographs are extraordinarily detailed and say as much about the human mind as they do about bird biology.

Back in the foyer of the conference hall, the lull in the conversation that followed Julie's innocent pronouncement is picked up by Stephen, who arrives and introduces himself.

“Hi there. I'm Stephen, the ornithologist at Ormsley Mere. How is everyone?”

A special type introduces themselves as an ornithologist at a bird watchers convention. But Stephen is a professional. People pay him to play with birds all day, which gives Stephen boundless confidence among other birders, for he knows they are jealous. They would all prefer to be him. He thinks he can flirt with all the single ladies just by eluding to his profession.

"Swell," Mary-Lou said as Bob nodded.

"Yeah, good, thanks," Julie added, uninterested in Steven's plumage.

"What did you think of Dr. Martin's talk? We were able to provide some of the specimens for that study. We have a large collection of eggs at the Mere."

This is why Stephen thinks he has it over all the other punters at the venue. He can collect as well as watch. His bona fide status as a scientist means he can obtain licenses to collect eggs and trap birds to ring them, even acquire specimens for collection just like they did back in the good old days. Acquire is a euphemism for kill, skin, and catalogue. Steven is one of the few who obtains the raw material for bird taxonomy. This is the process of description and classification and a means to decide what is in the world of varieties, subspecies, species, genera, family, order, and class. Aves for birds are the class, hence, aviary.

The process of taxonomy gave Mary-Lou her Mississippi Sandhill Crane, but she would be mortified that cranes died so that the taxonomist could measure and describe their critical attributes. Only that is the process. A detailed description of the attributes that are similar enough to group organisms together but different enough to make distinctions between them. It is a delicate art. Finding attributes—bone shape, plumage, beak colours—that are distinctive yet similar. Of course, many of the defining attributes that the birdwatchers admire, the calls, foraging behaviour and the flight pattern that gave Eric his white rump swift without full visual confirmation of the white rump, are not readily available to the taxonomist. He must rely on the physical attributes of the specimens in front of him. Until all material in collections can get a gene profile from genetic fingerprinting, the yes/no of the formal description defines the species.

Steven is immune to such sentiment. He has no problem skinning the specimen, putting the carcass in mild alkali to clean the bones, or even pulling eggs from nests to preserve them. It turns out that for most bird species, egg size, shape, and colour patterns can also be diagnostic of species.

As a visual collector, Eric is immediately jealous of the free-range Stephen has to add material to the collections. This was how it used to be when a gentleman could either go himself or sponsor a fellow to travel the world and send back trophies for the walls of the country house. When one is a collector, the more unusual the specimen, the less likely a rival has it in theirs. So, the hunt for the unusual is intense. Yet we owe these gentlemen collectors a great deal.

African Goshawk, Accipiter tachiro, is a common species in Southern Africa. These eggs are from the collection in the Natural History Museum of Zimbabwe in Bulawayo.

Alfred Russell Wallace (1823-1913) was remarkable, a gentleman in a time when such a description meant something and a traveller when getting from one town to another in England was a feat of endurance. In the mid-1800s, he covered the length of the Malay Archipelago with limited old money compared to his peers back home, albeit with wealth beyond belief to most Indigenous peoples he met.

Wallace was a naturalist and collector of exotic animal specimens caught or shot in the jungles and mountainsides at dozens of locations on remote islands. He made what money he could from selling preserved bird skins, mammal skeletons, and pinned insects to museums and his less adventurous fellow gentlemen collectors back in Europe. He was also an extraordinary observer, recalling almost immediately if a butterfly or beetle was new to his collection. Anyone who has dabbled in natural history collections will recognize this skill. A good naturalist needs to be proficient at it, and Wallace was, by all accounts, exceptional.

But what made Wallace truly remarkable was that he didn't just observe nature; he observed people, too. He saw the forms, behaviours and habits of all the locals he met on his travels. Indigenous peoples and colonists alike all fascinated him. Then, he tried to explain the patterns and behaviours of the people with the same logic he applied to nature. No doubt, he also carried with him the prejudices of his peers. His times were of colonial rule, the greatness of empire and knowing one's place in the pile. It is unlikely his observations were not clouded by at least some prejudice.

His seminal work, "*The Malay Archipelago*", is well worth its 500+ pages. At once, a primer on natural history, a window into thinking on the theory of evolution, a catalogue of the biology and peoples of the region, a gentleman's travelogue, a commentary on economics and an insight into how the Victorians saw the world. Often, the prose carries all these things in the space of a few pages.

Crossing the deep and treacherous straits between Bali and Lombok in 1859, he landed and began collecting, no doubt with an expectation of what he would find. Only a large number of species found on the western side of the strait were organisms related to Asiatic species, and on the eastern side, a mixture of species of Asian and Australian origin. Wallace saw in that line, later named the Wallace Line after him by Thomas Huxley, a disruption to the pattern he expected to see. The continuous variation that he followed in the myriad diversity of animals that fell into his traps should not have such an abrupt disjunct.

The genius of Wallace that he shared with Darwin was not only to recognise the disjunct, a sharp change in the frequency of new species as he crossed the straits but to realise it meant something profound. Gathering his observations together, he figured out what Darwin had also discovered: the concept of evolution by natural selection. His views diverged from Darwin's, but the two, through both private correspondence and published works, exchanged knowledge and stimulated each other's ideas and theories over an extended period. They both agreed on the importance of natural selection and some of the factors responsible for it: competition between species and geographical isolation. They diverged on why. Wallace gave evolution a purpose in maintaining species' fitness to their environment, whereas Darwin hesitated to attribute any purpose to a random natural process.

Although we now agree with Darwin on randomness, Wallace remained an ardent defender of natural selection all his life. And remember, this was highly contentious stuff at the time. In 1889, Wallace published the book Darwinism as a response to the scientific critics of natural selection. It was what a gentleman would do.

This gushing account of Alfred Wallace was written soon after I read the Malay Archipelago. I was and still am amazed by what some of the old-timers managed to get done and more to figure out what a bunch of it meant. They are inspirations.

I am struck by how little the gentlemen and women of today inspire me. Where are they? I have to assume that they exist. They do not seem to be among the celebrities or the internet influencers driven only by narcissism. Heaven help you if you try to find them in seats of government. They are not in the business community that continues to promote the economic model that is both precarious and does not give a jot about the resource base—much of the biology that Wallace saw is gone forever. They are not among the clergy. I cannot even bring myself to go there.

Perhaps, and here is a left-field assertion I never thought possible. Perhaps they are hiding among the scientists. I know, crazy suggestion, right? We are talking about the fence-sitting nerds who have bamboozled us for decades and failed miserably to get their messages across. Well, there is an idea promoted by Sir Paul Nurse, Nobel Laureate and chief executive of the Francis Crick Institute, that the response to the COVID-19 pandemic is to cause science to be done differently with collaboration, speed and efficiency up there as the biggest changes. In a crisis, the scientists work together to get through it as fast as possible while their fellow technical specialists, doctors, nurses and support staff on the front line provide heroic bravery. That sounds like what gentlemen would do, and, as we will soon see, it has plenty to do with the rare things.

Chapter Two

Millipedes

There has always been an argument over why *Homo sapiens* became so successful. Other stronger, more fierce species existed, even among the bipedal primates. Perhaps it was our big brains, language, adaptability, capture of fire, and even our belief in Gods. Often forgotten is the innate ability to seek and find the unusual. The instincts humans have to collect are more important than producing museums and galleries to visit or even the desire to classify objects. It comes from a survival tactic. We are exceptional at curiosity. And we add memory to the inquisitiveness that lets us know when we have seen something unusual. This skill was invaluable in the cut and thrust of the savanna, where every edible morsel was necessary to get through to the next feast on the kudu that a hunter managed to steal from a lone lioness. I kid you not. There is a theory that human stomachs are acidic, which enables the safe consumption of carrion. Getting some nutrition between the hunter's successes was vital—curiosity to find that difference. The leaf had a tuber underneath, and the red berry was a slightly different colour and not poisonous. The shape of the rock pool is likely to house a sizable crab. Whilst holes in the sand leak bubbles that house shellfish, not worms. We could figure these things out and then retain a search image for that foodstuff for a long time. We even figured out how to pass that on through the generations. Longevity is helped by women not continuing to reproduce beyond a certain time of life because giving birth is a huge, energetic cost that could otherwise be channelled into survival. That longevity helps retain the knowledge long enough to teach it to the young. That means to look out for the cicadas every seven years. Know that the termites will fly after the second big storm of the year or that spring is only when the forest flowers appear. Collections are an obvious extension of this curiosity in search image capability.

Major soldier and nymphs of the large fungus-growing termite Macrotermes michaelseni that is commen where it occurs in southern Africa

The genius of Darwin and Wallace and many of their fellow collectors was honed skills at spotting differences. The shape of the mandible of the soldier termite had a slightly more acute curve, and the colour of the beak of the Mew Gull was a little off in hue. Remarkably, they did this while allowing some natural variation between individuals but still knowing when a difference was big enough. These days, we get the computers to do all the heavy lifting. Those guys did it in their heads. And this is how it went for a long time. Collectors travelled far and wide and began overflowing the cabinets and corridors of museums and private houses with specimens of all creatures, great and small.

Everywhere collectors went, they encountered common species. In each group of organisms, there would be several types that would always be abundant. The types varied from habitat to habitat, but the pattern was always the same. There are a few common, abundant species and many rare ones, with a few in the middle. This pattern of rank abundance was intriguing, not least for its ubiquity. The theoreticians tried to explain it. They began with rank abundance curves, which rank all the species sampled in a habitat from the most to the least abundant—the same graph we saw in Chapter 1. The rank abundance curve visually depicts both species richness and species evenness. Species richness can be viewed as the number of species on the chart, i.e. how many species were ranked. Species evenness is reflected in the slope of the line that fits the graph. A steep gradient indicates low evenness, as the high-ranking species have much higher abundance than the low-ranking species. A shallow gradient indicates high evenness as the abundance of different species is similar. Here is an example of bird species diversity in a wetland habitat in Pakistan.

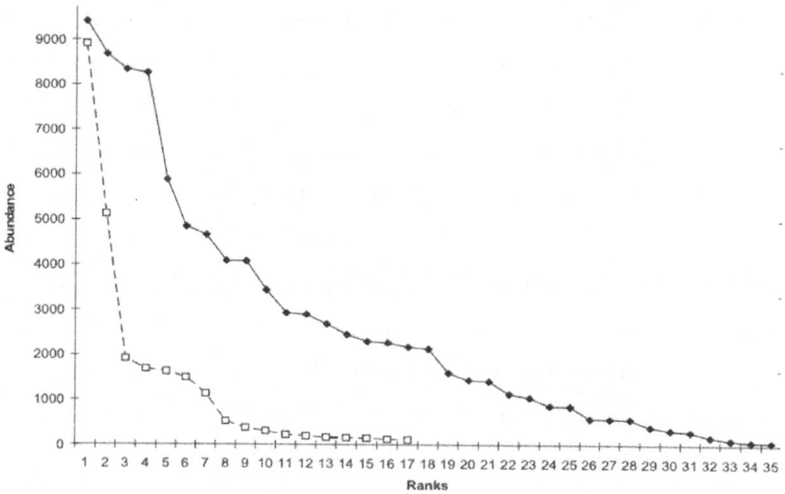

Species rank-abundance curves for the same habitat in different years. Source: Dauda, T. O., Baksh, M. H., & Shahrul, A. M. S. (2017). Birds' species diversity measurement of Uchali Wetland (Ramsar site) Pakistan. Journal of Asia-Pacific Biodiversity, 10(2), 167-174.

It's all good and exactly what we might expect from the eggheads. Lots of ideas and explanations, few of which seem to fit everywhere. This is typical for ecology and biodiversity specialists, especially. The question is why is there a rank abundance pattern at all. Why are some species abundant and others not so much? The explanation begins with a closer look at Charles Darwin's theory of natural selection. Here is the gist of it.

All organisms are different, and even those of the same species differ slightly. Darwin realised that variation made some individuals more efficient, less easily caught or in any number of ways more suited to the current conditions. They were favoured. What favoured meant was not some judgement passed by a diety or a grandiose human, it was simply a greater likelihood of survival and reproduction. Being better at life is more likely to make more of life. The individuals with the combination of attributes that gave them survival or energetic advantages over their fellows were more likely to produce offspring that survived. Over time and many generations, these often small survival advantages gradually paid off, and the best fit would increase in proportion in the local area at the expense of the less well-fit who would struggle to gather the same amount of resources or avoid the predator or frost. The survival of the best fit to the conditions was corrupted to survival of the fittest, which does not mean the most competitive or the most able to climb ladders; it means survival and reproduction of the best fit.

There are a couple of things to remember at this point. This process of natural selection never stops. It has been ongoing since the dawn of life on Earth and will continue even if a species comes along that can cheat it. The other thing to remember is that the best fit will change if the local conditions change. Seeds germinating early in the understory of deciduous woodlands have the best chance of basking in the sunlight before the trees put on leaves and shade out the sun. But seedlings are vulnerable to frost, so they germinate too early, and the last winter, Jack will get them—a delicate case of the best fit. Should the climate get warmer, the risk-takers will benefit, for there could be fewer deep frosts in early spring.

In order to influence survival relative to the attributes organisms possess, any change in local conditions has to be unusual relative to the organism's lifecycle. Weather events and disturbances with a consistent frequency would already be absorbed into the best-fit attributes of all the organisms in a given habitat. Some of the bigger sedentary ones even contribute to the attributes of the habitat, as do all the predators, diseases and parasites. The thing is that these changes are present as disturbances such as floods, fires, droughts or systematic trends, such as climate shifts and succession and erosion, the long-term breathing of the planet.

Natural selection favours the best fit, but today's bit best fit is not the same as tomorrow's. Rather than think of this in terms of new species or the process of evolution, let's consider the consequences another way. Suppose the best fit is to be flexible when the climate changes by being more tolerant of the conditions. Perhaps be a generalist forager or have seeds that germinate and produce robust seedlings in many different soil types. These species tolerate a wide range of conditions; they survive well and reproduce successfully even in dry times or during floods. They are adaptive and flexible. If natural selection works, then it can do so to favour the flexible just as easily as the specialist. Of course, the fit can also be precise. The best at being fertilised by pollen delivered by a specific kind of moth or a beak just the right shape to winkle out grubs in the tree bark. Natural selection can also make specialists.

Impalas are classed by the International Union for Conservation of Nature (IUCN) as least concern, with a stable population of around two million.

In the decade I lived in Africa, on every field trip, I would admire the sprightly and tenacious impala, *Aepyceros melampus*, often in large numbers and every day of the trip. The species ranges from Uganda, south through east Africa to Zambia, Zimbabwe, Botswana, Namibia, and South Africa. They occur in various habitats, can eat almost anything and are agile enough to evade attacks by predators, while excellent eyesight and group living help them see the predators coming. They are also prolific, able to reproduce rapidly, and so often persist in large numbers that can swamp local predators; they are the goats of the bush. Impala uses flexibility as the best fit.

Crocodiles are ancient creatures related to dinosaurs and birds with an evolutionary history dating back over 200 million years to the Mesozoic Era. They managed to survive the extinction event at the end of the Cretaceous period, about 66 million years ago, that took out the dinosaurs, yet kept many typical features that had been successful for such a long time. Modern crocodile species began to emerge and diversify during the Eocene epoch, around 55 to 34 million years ago. True crocodiles (Crocodylidae) started to take on forms more similar to what we recognize today around 37 million years ago. The Nile crocodile, *Crocodylus niloticus*, emerged around 16 to 20 million years ago during the Miocene epoch and the saltwater crocodile, *Crocodylus porosus*, dates back around 4 to 6 million years ago, during the Pliocene epoch. These are old species and are best fit because their lifestyle and body plan to enact it persist through disturbances that killed off specialists.

The Bracken fern, *Pteridium aquilinum,* is a global generalist species that thrives in temperate and subtropical climates in various habitats, including forests, grasslands, moorlands, and disturbed areas. It is especially resilient and readily colonizes open spaces. I am sure you have seen one. I can see a carpet of them from my office window. Fossil records suggest that ferns similar to the modern bracken fern have existed for over 55 million years, dating back to the Paleocene epoch. The species we recognize today, *Pteridium aquilinum*, has likely undergone evolutionary changes but has maintained its presence and characteristics over millions of years. Its best fit is as a generalist in both space and time.

Pteridium aquilinum, a fern species as bracken, brake, pasture brake, common bracken, and eagle fern is originally native to North America and Eurasia, the remarkably lightweight spores of this fern have allowed it to spread and establish a worldwide presence.

Like Wallace, Charles Darwin was a remarkable naturalist. Most would be more than satisfied with a great and revolutionary treatise, *On the Origin of Species by Means of Natural Selection, or The Preservation of Favoured Races in the Struggle for Life*, written over many years and first published in 1859. But Darwin wrote another seminal book, *The Descent of Man and Selection in Relation to Sex*, which describes another fundamental aspect of evolution. Sexual selection is a mode of natural selection in which members of one biological sex choose mates of the other sex to mate with (intersexual selection) and compete with members of the same sex for access to members of the opposite sex (intrasexual selection). These two selection forms mean that some individuals have greater reproductive success than others within a population, for example, because they are more attractive or prefer more attractive partners to produce offspring. Most evolutionary biologists accept sexual selection as a subset of natural selection, as Darwin no doubt intended.

Adult specimen of the common but rarely observed millipede species Alloporus uncinatus. They spend most of the year aestivating in the soil and emerge to feed and mate after significant rainfall events.

The mechanisms of best fit still apply, and the spoils were directly related to reproductive success. What is important for discussing rare things is that sexual selection can create species where there is otherwise considerable ecological overlap. This is not necessary, but it does happen. Species can evolve through mating, not just responses to survival and reproduction. While they proceed to do the same ecological function, they separate in the competition for mates. And we are not just talking about lions fighting for pride dominance or red deer on a lek. Sexual selection can pop up in any organism with sexual reproduction. Millipedes, for example.

These ubiquitous animals are tubes for processing various forms of detritus, dead organic matter. They shovel in leaf litter, soil, and algal crusts at one end, recover what nutrients they can and pellet out what's left at the other. They are throughput machines, and they all do it. As an enthusiastic young academic, I spent many hours researching their behaviours, activities, feeding, and mating tactics. A condensed version goes something like this, at least for the Spirostreptidae, the numerous tubes of interest that inhabit southern Africa's open woodlands and savanna habitats.

Males are thinner than females, have slightly longer legs and spend more time moving than females. Females are thicker than males partly to provide space for an elongated ovary with hundreds of eggs running alongside their tube-like gut. Females spend more time feeding to provide the eggs and so tend to be less mobile, often climbing trees to avoid the attention of males. Species can look similar, although they behave differently, but the taxonomists had an easy diagnostic tool. The seventh pair of legs in males are modified. First, we should say that millipedes, a thousand legs, don't have a thousand legs, but being Diplopoda, they have two pairs per body segment. It is an excellent arrangement for high torque and low-speed motion and is handy for burrowing into the soil to escape the long dry season. The seventh pair of legs in males are modified into gonopods, secondary genitalia. Not the organism of sperm production but the means by which to transfer sperm to the female as a key to a lock. Millipede gonopods are highly elaborate structures designed not only to fit the vulval lock but also to mix or even scoop out sperm that might be present in the spermathecae of the female, placed there by a previous rival. Being first doesn't win the prize, for each egg is fertilized as it passes through the spermathecae into the wild, one at a time. So when the males move around, hoping to collide with a female, they are sure they can get one over on the previous conquests.

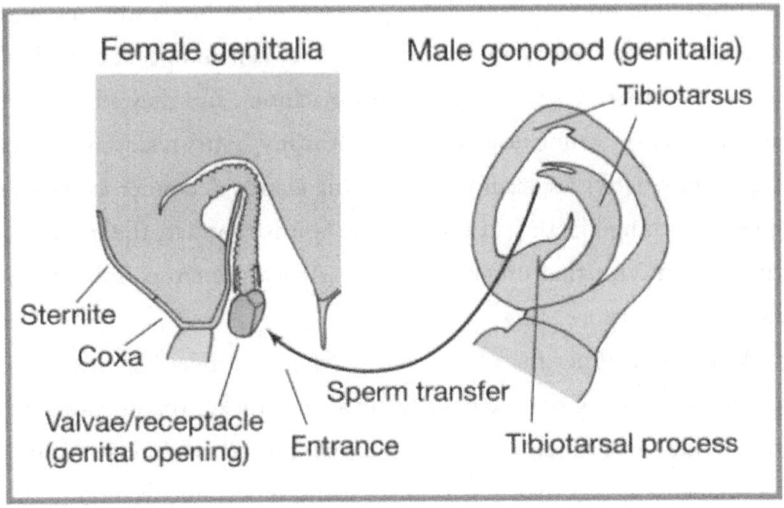

Drawing of millipede gonopods. These are extraordinary structures considering they are modified legs; they articulate and function to mix, scoop and transfer sperm and vary widely from species to species. Example from: Tanabe, T., & Sota, T. (2008). Complex copulatory behavior and the proximate effect of genital and body size differences on mechanical reproductive isolation in the millipede genus Parafontaria. The American Naturalist, 171(5), 692-699.

The only say the female seems to have is how elusive they can be. This takes two forms. Hiding, as mentioned earlier, or coiling into a tight spiral that takes some uncoiling beyond the strength of all but the biggest males. The males have another competitive trick. They can guard females they have mated with by extending copulation duration. They hold on for a lot longer if there are other males nearby. Makes sense for them. Wait long enough, and the rival males should move on to find a more available female. They can't hold on forever as both can't smoke the cigarette after sex for a day or both lose energy and for the male, other mating opportunities and for the female provisioning her eggs.

Sexual selection is a powerful evolutionary force. A species is just as likely to have evolved due to mate choice as selection for the best-fit feeding tactics thanks to a slightly longer beak. This means a lot for understanding rare things. They might be rare because they came about from some complex sexual competition. This is not how the story of rarity is portrayed. It is habitat loss or predation from an introduced predator or river pollution. Not always. It is possible, indeed probable, that rarity is the nature of things.

Most of my millipede research happened in Zimbabwe and Botswana during a nine-year stretch that took me to many wondrous places for memories etched onto my soul. The lion encounters were spectacular. As recounted in the preface, the first time, I stood on the same ground as a wild elephant. The smell of miombo woodland on a winter's morning. The broad smile of locals with only a few cents to their name. On my last day in Botswana, before I was due to leave for Australia, I walked along the dry riverbed of the Ngotwane River, which had been one of my millipede study sites. I sat on a fallen log to recall an amazing part of my life. I was sad to leave because Africa gets into your blood, even for white folk who know it is not ours. Also, there were pangs of excitement at a new adventure far away. I looked down as a soil biologist is instinctively drawn to do, and at my feet was a millipede. Cute, I thought. Only it was a little unusual. I picked it up, and sure enough, it was a new species to me. Nature has a sense of humour.

Most likely, it was also a new species to science. Not many myriapod taxonomists exist, and none worked in southern Africa then. As an ecologist, I lacked the tenacity and diligence of a taxonomist and relied on an instinctive eye for difference and published descriptions to name the species I worked on. That unusual specimen pitched up unexpectedly as a one-off most certainly eluded assiduous collectors who may have never passed that way. This is another critical issue in the story of the rare things. We only know what we know. If the catalogues of species are incomplete, then the number of rare types, no matter their origins, is unknown. If, as reported, there are 683 endangered or critically endangered bird species, does that mean there are hundreds more we don't know about? Birds we can see. Millipedes are a euphemistically 'poorly known' group of organisms. There's little information on any aspect of their biology, which is one reason my research on them passed the novelty test in the peer review process. So, there is an unknown number of species, a proportion of which are rare, and this could be high given the prevalence of the sexual selection mechanism. Many of them are critically endangered; we don't know which species they are because nobody has seen them, collected them or used taxonomic skills to describe them.

Millipedes are fascinating, but they are not charismatic. They do not have a royal society dedicated to their protection, and barely a handful of researchers have been able to spend time studying them. This lack of knowledge is true for most creepy crawlies. For every critically endangered bird or mammal, dozens of invertebrate species will be in just as precarious an evolutionary position. Many have faded away from the planet unrecorded by humanity. Nobody knows, and, it seems, very few people care. It is early in this epistle to dwell on this latter point, but for now, it is enough to note that humans do not record a significant proportion of the biological diversity in the world. As the millipede stories suggest, these small cryptic species are no less complex than the birds and the mammals, nor are they new.

Millipedes first appeared in the Silurian period over 440 million years ago, making them some of the oldest known land animals and giving them more than enough time to become hugely diverse. Some members of prehistoric groups were spectacular and grew to over two meters. These six-foot monsters were possible because of higher oxygen levels in the atmosphere. Arthropods rely on diffusion to get oxygen into their tissues, which limits their size in modern times. The largest modern species is the giant African millipede *Archiespirostrptus gigas*, which reaches 20 to 30 centimetres and is far heavier than the lightest mammal. Although size attracts our attention, bigger is better and all that, and the prospect of land invertebrates the length of a small car is the stuff of B-grade horror movies; that is not the takeaway from this part of the millipede story. Over 440 million years ago is the clue. That is a long time. The continents were in different locations, 100 million years before Pangea, the supercontinent that existed during the late and early Mesozoic eras. It assembled from earlier continental units approximately 355-335 million years ago and began to break apart about 175 million years ago. In contrast to the present earth and its distribution of continental mass, Pangea was centred on the equator and surrounded by the super ocean Panthalassa. Pangea is the most recent supercontinent to exist and the first to be reconstructed by geologists.

That intricate gonopod construction, the prevalence of sexual selection, and the appearance of rare species out of nowhere are less surprising when the lineage is this old yet well-represented. It suggests that the body plan and ecological role have persisted. Again, this is logical given that there has always been soil and decomposition to go with it. And with persistence comes opportunity for diversity through the mechanisms of natural selection, including selection in relation to sex. This introduces a hugley significant aspect of the how nature creates rarity. All that is needed for rarity to exist under the mechanisms of natural selection is time. Millipedes have had plenty of that, more than twice the time mammals have been around.

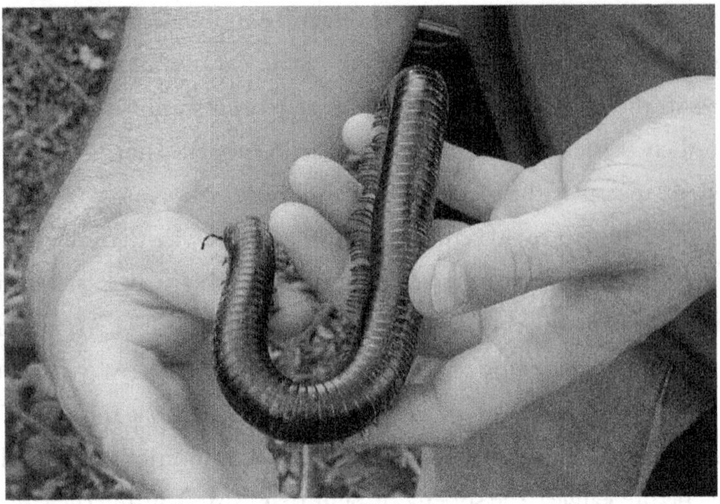

Giant millipede from the Mokolodi Game Reserve in Botswana but probably not Archiespirostrptus gigas, more likely a sibling species. No detailed studies of their gonopods, the key diagnostic character, exists to my knowledge.

Before I move us on, one more crucial attribute of rank-abundance patterns persists. There are a few common species and many that are less common—abundance is always unevenly allocated. It means that if there are ten species, perhaps three are common, three less common and three rare. One species is only seen on the last day of nine years of millipede research. But if there are 90 species, then ten species fall into this category of rarely, if ever, seen. Rarity is a numbers game. And diversity is about time.

Chapter Three

Reduction to near zero

According to the United Nations "*Determinants and consequences of population trends*", at the dawn of agriculture, about 8,000 BC, the world's population was around 5 million humans, roughly the current population of Sydney, Australia. Today, there are 78 cities around the world with larger populations. Tokyo is the largest city, with 37 million inhabitants, which is seven times the total human population before the dawn of agriculture. People were also widely spaced back then. There were no large centres or concentrations because there was no way to feed people who stayed still. Everyone had to persist in small groups around the resources they could exploit. We call this our hunter-gatherer phase because that sounds like a noble occupation. Brave hunters risk their lives to fell beasts much larger and more dangerous than themselves whilst their families forage for greens and edible tubers. Ah, yes, the noble savage. What a crock that is. Admittedly, *Homo sapiens* are smart, curious and co-ordinated with language to assist with plans and tactics. We were developing cultures to pass on information across the generations to give the young advantages that would be risky to learn over again. That seven-day-old carrion will kill you unless you cook it. There is always fresh water in the hill over the horizon where the sun comes from when it wakes. That sort of thing was a huge advantage over the bulk of the bear or the trotting power of the wolf. But the idea that nobility came from an ability to hunt when the reality was that back then, we were most likely clever opportunists and innovators with a knack for a search image. We were few and sparse on the landscape. The only large-scale impact we might have had, and we have to say 'might' because there is considerable conjecture over how important human impact was, is the loss of megaherbivores from the fauna when humans arrived. Before that story, though, we have to explain the one thing that our species has always been remarkable at achieving, and that is dispersal.

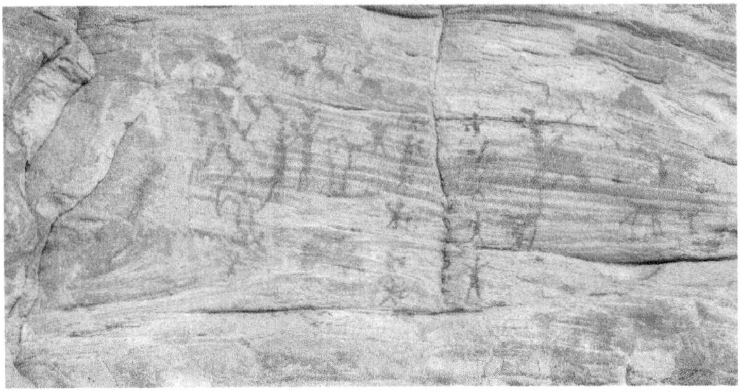

Modern humans, Homo sapiens, wanders across the world in small numbers for maybe 50,000 years before figuring out agriculture

The current thinking about our past is equivocal. The evidence from archaeology, genetics, and anthropology is that we began in Africa and stayed there for a long time, moving around the continent while other hominids competed with us, successfully for the most part. *Homo sapiens* are believed to have emerged in Africa about 300,000 years ago based in part on the thermoluminescence dating of artifacts and remains from

Jebel Irhoud, Morocco, published in 2017. The Florisbad skull from Floribad, South Africa, dated to about 259,000 years ago, has also been classified as early *Homo sapiens*. Previously the Omo remains excavated between 1967 and 1974 in Oman National Park, Ethiopia and dated to 200,000 years ago, were long held to be the oldest known fossils of *Homo sapiens*.

Whatever the origins, a commonly accepted pattern is the so-called "recent dispersal", which began about 70-50,000 years ago, a migration wave that led to the lasting spread of modern humans throughout the world. This means that most of the world before about 50,000 years ago had little or no influence from modern humans. Then, there was successful dispersal in a short period of evolutionary time. We went just about everywhere and persisted, albeit in low numbers.

Recall that the estimate is for 5 million of us by 12,000 years ago but widely dispersed across the continents. Then, some enterprising early humans in the fertile crescent of what is now Iran and Iraq and independently in China and South America figured out the basics of agriculture. Now, there are eight billion of us, three orders of magnitude more *H. sapiens* than in our glorious hunter-gatherer times. We will come to the causes of that population explosion shortly—it's not the invention of agriculture, by the way—but before we get there, try to imagine a world with just 5 million people and no roads, towns, or villages. There are no vehicles, coal mines or smelters. There is no countryside because we have not invented landscapes yet, just the natural mosaic of forest, savanna and grassy plains. In that natural habitat, there is **a greater diversity of plants, animals and microbes than at any other time since life began on the planet**.

Whenever the evolution of life on Earth is mentioned, there is great controversy over its origin—little green men visiting the planet, organic molecules arriving via asteroids, the almighty. A naturally fascinating unknown. We all would like to know where we came from. As you will guess, my pragmatic view is the replicator molecule emerged spontaneously in the organic soup of the pre-life planet, and what Darwin discovered took care of the rest. In college, I read *The Selfish Gene* by Richard Dawkins, the genetic mechanism for Darwin's natural selection. It explained everything and set me up with a paradigm to frame a career. So, I adopt the chance origin; no hand of God is required for seven days of exhausting creation and a genetic mechanism for selecting selfish genes. Along with this chance origin, I also accept that extinction happens because some combinations of genes do not persist. Genes might be selfish, but it does not guarantee that they persist; that is as much about chance as it being the best fit. The same applies to species that are specific and recognisable combinations of genes.

Why are species lost? Species go extinct because of the process that created them.

Ah ha, that's not the answer you were expecting. You thought I was going to say habitat loss, pollution, introduced pests, climate change or some other lefty topic sung about by old hippies and youngsters frightened for their future. Indulge me for a few minutes and follow this little sequence of logic to the end, and you'll see why extinction is more about the process of evolution than it is about humanity.

If you go to northern Manitoba, Canada, you can go to the wilds and see a polar bear. Tour operators will take you and look after you so the bears don't rip your Gortex jacket. You cannot see a wild polar bear in Australia. But, if you go out to the Australian bush, into the right habitat types, keep your eyes keen, and have some luck, you will see a koala. And, yes, a koala is not a bear; it is a marsupial mammal. There are no wild koalas in Canada.

Juxtapositions that are only possible with artificial intelligence

Canada has polar bears, and Australia has koalas. We know and take this for granted. The reason is that all species have regions where they occur in the wild and regions where they do not go and have never been seen, except in a zoo. It's called their geographic range.

Let's extend this idea a little. There are 828 bird species in Australia, give or take. Some are rare, others common. Some occur everywhere, and some only in one or two places or, in the case of the King Island Scrubtit, *Acanthornis magna greeniana,* on one island. It would take a lifetime of twitching holidays to see all these species because it is also natural for species to have certain habitats and places they prefer. Some species are super specialists and only occur in specific habitats, such as old-growth forests. As not much of this habitat is left, these species have a narrow range. You need to hike into the forest and be very patient to see one.

Other species are everywhere. The Indian Myna, *Acridotheres tristis,* or the House Sparrow, *Passer domesticus,* are familiar to half the people on the planet, even those who have never owned a pair of binoculars. Even in a habitat that occurs everywhere, like open woodland or backyards, not all species occur everywhere. A dozen or so species will frequent the average Aussie backyard. Species in the Melbourne suburbs will be different from those in Sydney, and in Brisbane, they will be different again. This is normal and always has been. Some species are rare; some are common. Some are localised, and others are everywhere. Welcome to nature.

In fact, half—400 or so—of the 828 bird species in Australia occur nowhere else on earth. They are endemic to the island continent or unique to it.

The House Sparrow, Passer domesticus, is an introduced species to Australia that now lives in most of eastern Australia and much of the Northern Territory and South Australia.

Alright, I've laboured this enough.

Evolution creates diversity as organisms focus on efficient ways to harness resources that they convert into more organisms; random mutation alongside natural selection results in diversity. If you replace organisms in the last sentence with genes, you get Richard Dawkins, Selfish Gene. I use organisms because a gene does not do it independently; it needs other genes to make a vehicle that can travel forward in intergenerational time. But, of course, Dawkins is right; genes are selected.

We are so familiar with the result of natural selection—the huge variety of species unevenly spread across regions and habitats—that we take it as given. Don't be stupid; everybody knows there are no wild polar bears in Australia; it's too hot for them. Nature comprises gene combinations that are the most efficient at converting the resources in a given place, given the conditions. This is what evolution does all the time, everywhere.

This is the bit you need to follow... **given the conditions**.

Specialisation is an inevitable result of natural selection. Crocodiles, for example, are still generalists with a fine working model of existence that has not changed much, but most species have become quite specialised. The problem with specialisation is what to do when the conditions change. Just ask the dinosaurs how that went for them. Again, we accept that mass extinction occurred 66 million years ago without blinking. A huge, diverse group of land animals died out quickly because a meteorite hit the earth and changed the conditions. Dinosaurs couldn't cope with what the impact and its aftermath did to their food and shelter. It disrupted things too much for them to adapt fast enough, except for the crocodiles, who could switch to fish.

The fifth mass extinction, which killed the dinosaurs and 76% of the species alive at the time, is a dramatic, iconic example of change. However, conditions change all the time, which means that species go extinct all the time. It is an inevitable part of the process.

I'll say this again. If conditions change and they change fast enough, acutely enough or for long enough, some species cannot survive and go extinct. Period. The flip side of change that causes extinction is the opportunity for diversity. Change things and gaps open up for new species to exploit. But we'll save that idea for later. Here we are on the extinction consequence.

Change is inevitable, and extinction is inevitable. However, this is not what we are told. Take this example of propaganda from the front page of the Guardian online feed on 31 March 2019, which states that record numbers of Australian mammals face 'imminent extinction' and includes this quote.

> *The report calls on the next Australian government to commit to a policy that no more of Australia's unique wildlife species will be allowed to go extinct, and to ensure, backed by the necessary laws and resources, that this policy is fulfilled*
> Peter Roberston, Wilderness Society's national forest campaigner.

Pause and read that quote again. This person is asking you to believe that extinction, a law of nature that has a history as long and deep as life itself, can be overturned by muppets in Canberra. Come on, get over yourselves. You are not that important. Human activity is the biggest current agent of change, but to think that a specific political decision can rewrite evolution means you do not understand the challenge.

Thinking we can 'save the koala' forces us into a bind. We now imagine that extinction is terrible, an awful result that must be prevented. So important is this premise we even allow politicians to spruik an impossible policy to stop it from happening. A very different view of conservation would appear if extinction were an acceptable option, one that happens in nature. Then, the choice to save a species from extinction would be a conscious one made with a clear understanding of human values. We may still want to save the koala, panda, or Southern Corroboree frog, but it's a conscious choice if we do.

Few people see a wild koala but the obsession with cuteness forced me to write 'The Mindful Sceptic Guide to Saving the Koala' debunking the myths

Extinction is inevitable and not just for species. Mass extinction is the next thing that crops up in the evolution of life. Just the term is enough to wreak havoc with the mind. Whole branches of the taxonomic tree heaved off and cast aside by environmental changes: loss, loss, and more loss.

The first big one was the Ordovician-Silurian extinction 444 million years ago, the second worst mass extinction known to science, which killed an estimated 85% of all species. The event took its hardest toll on marine organisms such as corals, shelled brachiopods, eel-like creatures called conodonts and the trilobites. All because the earth cooled and messed up with sea levels and the basic chemistry of the ocean waters. The next big one was the late Devonian 383 to 359 million years ago, which eliminated three-quarters of the species on the planet at the time. This event took 20 million years, and the causes are obscure perhaps volcanism. After the Devonian disaster came the 'great dying', the most devastating extinction event in the evolutionary record, and no, it's not the dinosaurs that comes later. Some 252 million years ago, life on Earth faced the 'Great Dying': the Permian-Triassic extinction. Over 60,000 years, 96 per cent of all marine species and about three of every four species on land died out. The world's forests were wiped out and didn't come back in force until about 10 million years later. Of the five mass extinctions, the Permian-Triassic is the only one that wiped out many insect species. Marine ecosystems took four to eight million years to recover. The extinction's single biggest cause was volcanic activity that erupted more than 720,000 cubic miles of lava across what is now Siberia, triggering the release of at least 14.5 trillion tons of carbon, more than 2.5 times the amount unleashed if every last ounce of fossil fuel on Earth were dug up and burned. Adding insult to injury, magma from the Siberian Traps infiltrated coal basins on its way toward the surface, probably releasing even more greenhouse gases such as methane.

At the end of the Triassic 201 million years ago, the earth warmed again, with CO_2 released due to a huge volcanic eruption in the ancient continent of Pangea. As already described, natural selection is great at delivering diversity so long as disturbances are within bounds and reasonably predictable. Anything out of the ordinary is challenging and potentially catastrophic if the changes are big enough. A global warming event of four to six degrees centigrade resulted in the loss of 80 per cent of land and marine species in existence at the time. Finally, we get to the famous Cretaceous-palaeocene extinction 66 million years ago, the most recent event that wiped out the dinosaurs and three-quarters of all species on the planet. The evidence points to an asteroid around 12 kilometres across hitting the Yakutan peninsula at ridiculous speed, forming a crater 190 kilometres wide and messing with the atmosphere something rotten, certainly out of the ordinary.

Captivating as these anecdotes of reduction to near zero may be, they also tap into our mortality and fear of death or play on that collector mentality. When stripped of their collection, perhaps in a fire or by an especially cruel thief, a collector of anything—Pokemon cards, old woodworking tools, little China plates, spoons with place names on them—is gutted. The loss of so much life wreaks great pain; so much biodiversity loss is unthinkable. The only reason we are okay with the loss of the charismatic dinosaurs is that some of them were huge and scary.

Across the evolutionary record, the talk is always extinction, usually mass extinction. A mentality of disaster. So here is a thing. Those five million modern humans wandering around the plant merrily hunting, gathering, and nicking lion kills when the great beasts were not looking did so in **the most biodiverse period in Earth's history**.

The tales of mass extinction get the headlines, as does the sixth mass extinction humans are causing right now, but what happened after each of these events is that life became more diverse. The peak of that diversity was just before we invented agriculture. When those five million souls were wandering, they were surrounded by more species than at any other time in the evolution of life on Earth. And given what we learned about rank abundance, these good folk shared the planet with more rare species than anyone could shake a fist at.

Here is the pattern for the Phanerozoic eon, the most recent 542 million years of Earth's history characterized by abundant and diverse life forms, including the evolution and proliferation of complex multi-cellular organisms.

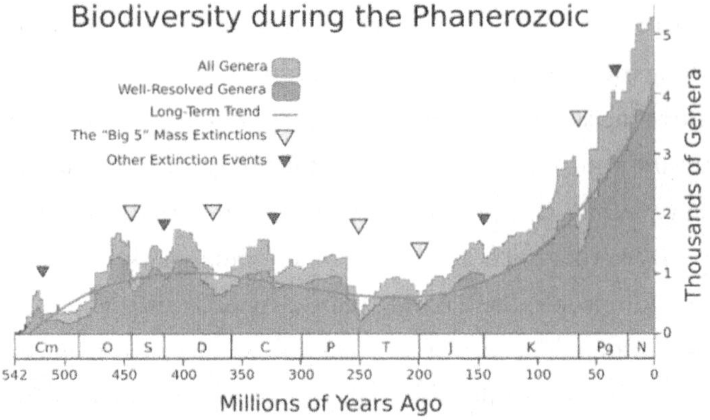

Despite numerous extinction events, including mass extinctions with the loss of over 75% of species, diversity has increased over evolutionary time. Image from Wikimedia Commons (CC BY-SA 3.0)

Extinction events get the headlines, but the process of natural selection can't stand a vacuum and rapidly recreates whatever diversity is lost. Of course, this does not negate the so-called crisis of biodiversity loss 8 billion humans, their livestock, and pets are causing. It will be millions of years before the diversity recovers to exceed the previous levels. But recovery will be possible, even if humans leave a wasteland behind. After 4 billion years, nature has so much raw material to work with that diversity recovery is inevitable.

True, *Homo sapiens* will not be around to see the fulfilment of it given the timeframe for the recovery and our likely demise from excess, but nature, gaia, the planet or whatever moniker applied to the soul of the earth couldn't care less. There is no moral or emotional underbelly of evolution; it's just an unstoppable process now that DNA-based life exists.

Some 650 million years ago, just before the pre-Cambrian explosion of multicellular life at the beginning of the diversity curve, there's a hypothesis that the earth was a gigantic ball of ice. It is a near-total extinction event. The Snowball Earth hypothesis proposes that during one or more of Earth's icehouse climates, the Earth's surface became entirely or nearly entirely frozen, sometime earlier than 650 Mya (million years ago) during the Cryogenian period. Proponents of the hypothesis argue that it best explains sedimentary deposits generally regarded as of glacial origin at tropical palaeolatitudes and other enigmatic features in the geological record. Opponents of the hypothesis contest the implications of the geological evidence for global glaciation and the geophysical feasibility of ice- or slush-covered ocean and emphasize the difficulty of escaping an all-frozen condition. Several unanswered questions remain, including whether the Earth was a full snowball or a "slushball" with a thin equatorial band of open, or seasonally open, water. We do know for sure that however extreme it got, life persisted. Microbes survived this period, perhaps in the slush or the water beneath it. That was enough to seed the radiations of the future.

There are a couple of important caveats to all this. The first is the paucity of data to generate these explanations. There are multiple sources of evidence from stratigraphy, geological formations, dating methods, fossil records, and evolution models. Still, the explanation that God did it all in a week is certainly more straightforward. Some of the claims in this chapter, and even the whole book, are likely wrong. Many are hypotheses with little or no substantive data to support them. Others are more reliable. Either way, dismissing or arguing with any of the specifics is easy. Academics have and will continue to do so long after we retire to our armchairs. And that is okay for the primary thesis, which is that **the immediate matters to humans but not to the grand scheme of things**. Gaia could not care less about extinction. She knows that there is plenty of time for recovery.

The second less cynical caveat is that we operate and think in the immediate. Can you remember what you had for dinner last Wednesday or on this date a year ago? It's unlikely unless Wednesday is always pesto pasta day. Our minds can project forward in an ecological time of months to decades, and we have some memories from the past on those scales, but we cannot even comprehend a million years. It is very hard to understand what evolution through natural selection can accomplish given that time.

Although long time frames elude us, nature is built on them. The pulses of the climate, the languid march of ecological succession, the slow and steady erosion of mountains and the building of soil and sediment. Diversity is inevitable in such a slow pace of change. There is more than enough time for organisms to adapt and for the genetic mechanisms of intergenerational transfer to hold on to the variety. The staggering amount of time between mass extinction events, typically over 50 million years and as much as 150 million years, is millions of generations for even the longest-lived organisms, an unfathomable amount of time for random variation and the fixing of preferred types. And also for their loss. On these timescales, the number of species going extinct is huge. The complete turnover of species in persistent habitats, like the African savannas, occurs often. Given enough time, extinction is inevitable.

Then, there is the extraordinary array of habitats. Caves and underground springs, salt lakes, vast rivers where one bank is obscured from the other, shallow seas, deep seas, mountain ranges, grassy woodlands, stony deserts, the list appears endless. This creates the opportunity for any number of resources, habitats and niches, as the ecologists call them.

That is a resource envelope that individuals of a given species prefer and are best suited to exploit. A species of frog is found only under the lip of Victoria Falls and in the wet forest in the spray above the gorge, a unique species of woodlouse *Aphiloscia victoriana*. This means species can be generated in time and in unique habitats. This allopatric speciation also referred to as geographic speciation, vicariant speciation, or its earlier name, the dumbbell model, is a mode of speciation that occurs when biological populations become geographically isolated from each other to an extent that prevents or interferes with gene flow. Classic examples of this are mountain ranges separating lowland specialists or the retreat of cold climate habitats to the tops of isolated mountains during warm periods. It also works if the habitat conditions are unique, such as a permanently moist layer of leaf litter in the woodland that normally dries out during the long dry season and is just a few hundred meters away from the spray plume.

Vigilant impala alongside the common tsessebe or sassaby (Damaliscus lunatus lunatus) is the southern, nominate subspecies of Damaliscus lunatus, although some authorities have recognised it as an independent species

The world that the five million humans inhabited before agriculture was full of biology. Diversity was everywhere, thanks to time and the myriad of opportunities for specialisation. Add to this speciation through sexual selection and competition for food or shelter, the mechanisms of sympatric speciation, and it was a truly biodiverse place.

Do not be fooled by all the talk of extinction either in the mass events of the geological past or the current mass event, the first endogenic one made by a single species on the others. All the talk of loss does not mean that diversity is extinguished. Indeed, the dramatic disturbances create evolutionary opportunity. From them comes radiation of types not before seen. We will lament the loss, for it is our nature to do so. But loss creates opportunity in an evolutionary story that is as much about extinction as it is about speciation.

Sir David Attenborough was born on May 8, 1926. At that time, there were just shy of 2 billion (2,000,000,000) people on Earth. Attenborough's documentary series "Life" (2009) was estimated to have been watched by over 500 million people worldwide. His series "Planet Earth" (2006) was seen by around 100 million people in the UK alone and broadcast in over 130 countries. "Blue Planet II" (2017) was watched by more than 14 million people in the UK on its premiere, making it the most-watched TV program of the year. Perhaps a billion people have seen some footage of the natural world that Sir David has brought to the masses. His is a long and remarkable life of travel and observation of nature in all habitats. His engaging style and genuine fascination for the variety of life are infectious. In person, he is just as passionate and committed to the joys of the natural world. Only in recent years, there has been a change. Not because of his advanced age, for he remains as sharp as a tack. What is different is that you can see he is frightened. Not of death for himself, but he is scared for the future of so many of the wild places he has seen. As he goes to them for the third and fourth time, often several decades after his first visit, the landscapes have changed. There are fewer creatures, all harder to find and film. Of all the people alive, Sir David knows that the sixth mass extinction is upon us.

This is sad. In our lifetimes, thousands of species will become extinct, either completely or functionally. This human generation will see more species extinction than any other. There will be iconic losses, creatures that many people know. Then there are other creatures that not even the specialist knows about, species that are lost from the earth before humans have described them. We don't know how many because the catalogues are incomplete.

Some of the iconics we could save. Resources can be used to rewild some of these species. There will be triage, though some will not be able to be rescued, and they will pass. These are challenging choices. Is a pardalote worth saving, the rhinoceros, the Andean condor, the Yellow-footed Skink, a butterfly in the heart of the Amazon that only the first peoples have ever seen? Surely not the scorpions, sea snakes, or mosquito species that transmit parasites. Perhaps we should save the rare millipede species in the fringing woodland of the Ngotwane River in Botswana.

We will see the reason for this mass extinction soon, but there is another important feature for now. What happens if half the individuals in a population of skinks die of a mysterious disease? Trick question because the answer depends on the size of the population. Not the absolute size but the effective size. Imagine that you are a male skink. In an evolutionary sense, you have only one primary task: to find a female skink and fertilize as many of her eggs as you can. Crude but true. As you wander in search of success, the reduction in numbers kicks in. Your search takes longer than it did last year. Indeed, it has been many weeks, and there are no females anywhere. This is a problem. The breeding season will be over shortly, and there is a winter to get through before the next one. Needless to say, you will keep looking. If the original population of skinks were small, this long search was normal, but this time, the loss of half the already scattered opportunities means failure. The same applies to the females; they risk not being found by a male. Overall, the population declines further when males do not find females. The skink will go from rare to close to extinction in a few generations. The loss of half can be terminal. Of course, if you compete with other skinks for food, reducing their numbers is a bonus for the survivors. Food should become more plentiful, increasing chances of survival and providing spare energy to look for a mate. What matters is not how many are lost or even how many are left behind but how numbers relate to their critically limiting resources or resources that determine the evolutionary fate of individuals, populations, and species.

Every place that Sir David revisits as a 90-year-old has changed since the first time he visited. His career began as a series of quests where he honed his unique style of bringing wild creatures into people's living rooms. He is probably one of the most travelled individuals alive; be it deforestation in Madagascar, tourism in the Galapagos, agricultural expansion into the Amazon or palm oil plantations in Indonesia, he has seen the changes. He said little about it for years, certainly not in his films. They were educational and entertaining, not intended for genres of disaster or horror. Only in the twilight of his career did he start to tell it as it is. Land clearing, pollution, agricultural expansion, poaching and climate change were all culprits.

The common agama, red-headed rock agama or rainbow agama (Agama agama) is a species of lizard from the family Agamidae found in most of sub-Saharan Africa. This one I photographed near the grave of Cecil Rhodes in Matopos, Zimbabwe

Check out any of Sir David's recent awards speeches on YouTube, especially those for the Earthshot prize ceremonies. When you look closely past the glitteringly sharp eyes and the carefully scripted words, there is a sadness. He went to places resembling the Earth of those five million souls who wandered around before agriculture. No fences, sharp edges, monocultures, hedgerows, drystone walls, or treeless fields that disappear over the horizon. Just plants living where and how they could, all intermingled and wild. He filmed the large animals alongside microbes and tiny animals in the soil, on leaves, and symbiotic with plant roots. The diversity of life, as Ed Wilson's book title, gathers it all up—the consequence of all those millennia of natural selection and more making. Most pristine places are gone or irrevocably altered because of a global change big enough to herald a new geological epoch.

We can't put it off any longer. We have to take on the Anthropocene.

Chapter Four

The Anthropocene

A subdivision of geological time does not happen every day. Only a handful are recognised from the billions of years of history recorded in the earth's rock layers. Consistent features must mark subdivisions in the stratigraphic record and clear breaks that would separate one time period from another on some logical grounds. The boffins on the subject are still thinking about it for the Anthropocene. While not officially recognized as a distinct subdivision of geologic time by the International Commission on Stratigraphy (ICS) or the International Union of Geological Sciences (IUGS) as of July 2020, the Anthropocene Working Group (AWG) of the ICS's Subcommission on Quaternary Stratigraphy (SQS) has taken steps toward its formal designation. They are unsure that what humans have done to the planet will leave an indelible mark on the rock layers. It will. For the first time, a single species, not some wild climate effect or extraterrestrial body, has caused a distinct break in the geological record.

There is a weird schizophrenia among those fond of this new geological epoch. Excitement over witnessing such an event is palpable. No geologist can pass up the significance. Then there is the name Anthropocene. We did it, and it felt like we were making a speech after a once-popular actor gave us a golden statue. How many people can I thank? I hope my mom's watching. Then there is the truth of the matter. Heaven help us, what have we done? Don't they always say, don't shit where you eat. Proud but scared. What an amazing mess that is, and this explains some of it.

> "replenish the earth, and subdue it: and have dominion over the fish of the sea, and over the fowl of the air, and over every living thing that moveth upon the earth".
>
> Genesis 1:28

An interesting injunction, isn't it, to have dominion? One that explains a lot about human behaviour, noting that the behaviours were honed long before religious instruction. It is pertinent to those people in the West, many of them lapsed churchgoers who still default to the word of the Lord. The edict sets up a war on nature because if we don't fight the wilderness, it will consume us, and the evil forces of entropy will turn us into dust. We must bring nature under our control by capturing its production for our food and fibre and removing the nasty predators and pests from our presence—all about me.

It wasn't always like this. Back before agriculture, humans were few and lacked the power for dominion. We are told that early people approached nature and did things very differently, even as they changed nature for their benefit. More than once, I have stood in a rainforest that modern science considers pristine. But the first peoples who lived there knew it was moulded by their ancestors, who planted what they wanted and discouraged species that had no obvious use. The result of millennia of human interventions is that the proportion of 'useful' trees is far greater than by chance because people increased the percentage of useful ones. The same applies to the landscapes of Australia. There is little doubt that the use of fire by aboriginal peoples gave the continent its eucalyptus forests dominated by tree species able to regenerate from seed or epicormic growth even after severe fires. Then, the western farmers came to cut down the trees.

A self-seeded Norfolk Pine, Araucaria heterophylla, growing on the shoreline behind the Tuross Head Memorial Walls, NSW South Coast, Australia.

Manipulation of nature is older than the plough and much older than the bulldozer and the tractor. However, we could borrow from these older, more benign ways of moulding nature that are more collective than for the individual. A forest dweller who sets a tree seedling to create a forest 'garden' close to his village will not see the fruits of that tree. His son might even need to wait until he is an elder. Collective use was intergenerational care, not something for today's market. But I digress.

Dominion sits with our need to make more, that primal drive from our genes. And because humans are exceptional at more making, we can keep our young alive barring accident and don't feel the need to have a dozen kids anymore. We no longer die young from lack of food, exhaustion or failed reproduction. Have dominion, we told ourselves, and so we have, all the way to a new epoch.

Nearly all the evil antagonists in films and TV dramas start the movie with mansions, large cars, a battery of heavies and beautiful women. Convention has it that they are men because convention also has it that it is a man's world. Hunting is braver, stronger, and more domineering than gathering with an infant strapped to your back. The evil crime boss is wealthy, wallows in the trappings of it all and always comes a cropper. The shabby and flawed detective eventually figures it out, risking everything. Part of us is always a bit jealous of the villain. The trappings of dominion smell like freshly baked cinnamon scrolls that take some resisting. Luckily, our morality is rescued by the flawed protagonist, who is prepared to work all hours, get beaten up, ignore his loved ones, and do whatever it takes to get justice done—our hero.

Dominion over the earth makes us the villain in the analogy, wallowing contentedly in the spoils. Extinction is that detective, slowly but surely, bringing us down.

Sir David Attenborough has seen the consequences of change from habitat loss, pollution events, the profusion of invasive species, and poaching. There is a single cause for all of these: humanity. We have followed our biology and made more; only we are so good at it that the consequences have gotten out of hand. We grow enough food to have healthy women able to reproduce at near maximum rate, unless they choose not to, and have reduced infant and adult mortality. One metric of improved survival is life expectancy. Through the 1700s and 1800s, it was 28 years, and then during the initial phase of the Industrial Revolution from 1900 to 1950, it jumped from 32 to 47 years. In my lifetime, from 1961 to 2023, it has risen steadily from 50 to 71 years. A child born today has a good chance of living a long time.

So, as fertility rates decline, it's mostly because we choose to have the number of children we want. This is not a book about demography or our dominion, so no more whinging and lamenting the loss. Just look at Sir David's eyes when he talks about the extinction crisis; all you need to know is written there. Those eyes have seen it, the beauty and the beast.

Let's accept that dominion has happened. Humans now control a quarter of the global net primary production—15.6 Pg C/yr or 23.8% of potential net primary productivity, of which 53% was contributed by harvest, 40% by land-use-induced productivity changes, and 7% by human-induced fires. We have touched every square foot of the land and dumped garbage into the ocean's depths. It is what it is.

Suggestions and arguments exist to limit the damage from 8 billion and counting. There are shelves of books on the topic. In the meantime, what does dominion mean for the rare things? In short, there is a much greater likelihood that the already rare will disappear altogether, and there will be an increase in the newly rare. There is more extinction than before, and more species are becoming rare. That is what mass extinction means: loss of the masses. The number of species lost will be higher than at any other time in Earth's history because the extinction event began at the most biodiverse time when evolution had made the most richness. The proportional loss is a subject for conjecture but will be similar to past events; perhaps eight out of ten species lost.

Before we look at the mechanisms more closely, there is another reality to take on board. There is nothing we can do about it. This extinction event is already in progress. We may have some options to limit the depth of the impact, perhaps get to seven out of ten, and we might also have the resources and the will to save some iconic species. But your average staphylinid beetle or diplopod or dipteran are unlikely to gain our attention. There will be more and more rare things, and very little we can do to keep them from extinction.

Depressing. Sobering too. A species with a brain capable of morality and imagining its own death is responsible for the loss of billions of sentient beings while putting its future in doubt. More than the rare things indeed.

This is the Hong Kong skyline where more people live above than below the 14th floor of tower blocks.

When we looked briefly at the previous mass extinctions, the causes were mostly climate-related. A cooling in the Ordovician, oxygen level reductions in the Devonian, the release of carbon from lava flows in the Permian causing a warming event, another warming event in the Triassic probably from magma in the centre of Pangea, and then the famous meteorite strike in the Cretaceous 66 million years ago pushing up enough dust to cloud the atmosphere to cool the planet rapidly. Climate change is the primary culprit of mass extinctions, most often a change in the proportion of atmospheric gases, but also from too much dust. This is one of the reasons that it is important to pay attention to human-induced climate change. Shifts in CO_2, oxygen, and greenhouse gases like methane have been catastrophic in the past and can be again.

Evolution likes disturbance, but the species it generates not so much if the intensity or duration of the changes are outside normal tolerance. A dry year or two is not so bad. There will be wet refugia for the species that can't tolerate dusty soil. The change that keeps it dry for decades and causes refugia to dry up, too.

A good friend of mine owns a plot of land inland from the coast in the east of NSW, about half an hour's drive outside the small town of Taree. He grows a few tree crops, but most of his land and that surrounding it is forest, officially wet sclerophyll forest. The key part is wet. Two streams run permanently through his place. Chris also has a small nursery on the farm, as for many years, he has run a successful bush regeneration business, helping local landholders rehabilitate and rejuvenate degraded bushland. He knows his vegetation, what grows, where, when, and how well, what the fire frequencies are, and what happens after a fire. Since the mid-1980s, when he bought the block, several fires have blazed close by in the dryer country on the hills, but never down into his wet forest. The plant species present, their size and conditions, suggesting that this vegetation has not burnt for at least a hundred years, perhaps a lot longer. In the Black Summer of 2019, fires from the west and north converged on the wet forest. It was dry enough to burn through all the vegetation and destroy all the outbuildings on Chris' farm. Luckily, my friend was on holiday; otherwise, he might have been in harm's way, trying to beat back the flames. It's hard to do that with large wildfires, even with aircraft and fire tenders. Remote bush blocks are saved only by the wet habitat unless the drought is so deep there is not enough moisture to slow the fire.

As a keen observer of native vegetation, Chris has watched the recovery of his wet gully habitats as the rains have returned and the streams are flowing again. He concludes that the wet sclerophyll forest will never be the same again. The area would have burned in the past, perhaps two or three hundred years ago. Still, there was no lantana (*Lantana camara),* an especially invasive weed from Central and South America, in previous burning events. Even if the natives persist, they must wait their turn while this shrubby tree is momentarily in the limelight.

Lantana camara is a heavily branched shrub that can grow in compact clumps, dense thickets or as a climbing vine and can smother native vegetation forming impenetrable stands. Native to the tropical and subtropical regions of North, Central and South America, Lantana camara now covers 5 million hectares throughout most coastal and hinterland areas of Australia.

Many will claim that the sixth mass extinction results from human-induced climate change. This version is handy because we are arrogant enough to think we can reverse that catastrophe and return the climate to normal. We should reduce emissions, for failing to do so only worsens the problem. But even if the coal-fired power stations closed, the cars are phased out for their electric or hydrogen fuel cell cousins, and pandemics reduce our wanderlust the climate has already changed. The drought that preceded the 2019-20 bushfire season in eastern Australia was deep and long. Some locations had not seen rain for four years. The temperatures were some of the highest ever, with the 2010s being the hottest decade on record. It is here and is biting. One of Donald Trump's best gaffs was during a cold snap in 2019 when a wintry blast from the Arctic stretched further south than usual. Where is your global warming when we need it? Well, Mr. President, all that extra energy trapped in the atmosphere does not have to wait for summer to do its thing; as your primary school teacher told you, sir, heat is energy. At the end of this chapter are some comments on climate change, for this is a consuming topic. Anything that takes time to emerge is chronic rather than acute and cannot be stopped but still requires long-term plans to adapt. It is everything the modern politician hates. Climate is bigger than him, out of his control, and has its own narrative. It is unfixable and yet will still cost money and lives. No wonder the populists double down on the lies. The geological record will have climate as the cause of the sixth mass extinction at the end of the Pleistocene, listing the cause as human. But all the extra greenhouse gas was the ultimate cause, not the proximate one. To discover the real reason for the sixth mass extinction heralding the Anthropocene, we must start 12,000 years ago, when humans invented agriculture.

As with all things prehistory, there is conjecture over how, where, and when the first crops were grown by hominids. Whatever the beginning, there's no doubt about the why—dominion and power. A food supply that lasted all year around and was sufficient to feed more than just a few in the household became the means to control territory and to defend it. Enough food meant place. A patch of land worth defending from all comers, worth building shelters on that were strong, warm and weatherproof. No more winters in the cold huddled around fires. No more children snatched by lions or wolves in the night. All year-round food for more than a few was a total game changer. There can be debate about when the first agriculture began and who invented it. There were multiple instances and no doubt these questions will be argued over. What should not be ignored is the game-changing. Cultures were created, and even though some, such as Great Zimbabwe, the Inca culture and the Easter Island peoples, collapsed, the idea didn't. Today, agriculture can support a city of 35 million people.

If it is possible to imagine the invention of agriculture, then here is one possible scenario. Food is sometimes scarce, forcing people to exist in small groups. This is fine enough against individual predators and other small groups but lacks stability. There is always the threat of starvation and the need to stay nimble and mobile should food or water run out. Then, some bright spark threw away a few grain seeds; maybe a gourd broke after a foraging trip. The seeds germinate, and after some time, many seed heads emerge right next to the crude shelters. Do it again, only more of it. When we return from our seasonal forage to the coast, there will be grains for us. Imagine the esteem such individuals would have. Food will be there when we get back. And it was. What sort of magician is this? Let's put more effort into this miracle and scatter more grain than we eat. Ah, but they didn't grow so well among the trees. That is not a problem. We can chop the trees down with this stone axe.

It might have taken a thousand years for the early farmers to start altering habitats to suit better the new crops they cultivated, maybe a lot longer, but this revolution in how humans lived had two dramatic effects. It altered the landscapes and dramatically increased the number of humans alive to around 160 million. This represented a two-order of magnitude increase in the number of mouths to feed. Still, after 10,000 years of figuring it all out, there was an infrastructure of production sufficient to run the Roman Empire and the Greek civilization before that.

Large tracks of land were converted to food production. What that is in ecological terms is a channelling of primary production, plant biomass, into a subset of plant species that have edible parts for humans. Grains, fruits and tubers predominantly from both annual and perennial plants. Recall that the diversity in that original biomass, the forests, woodlands, and grasslands, was as high as ever. Agricultural practice simplified biomass production to just a handful of the most useful species. This was a lengthy process of simplification. In the early days of agriculture, the plants were wild. Many different types would be used for each area's best option. One group of newly settled humans in south-central Mexico hit the jackpot. Their local grain is *Zea mays*, the species that produces nutritious cobs, which are small but easy to use. After a while, they figure they can sell the surplus of this desirable plant. The plants improved in value by noticing that the biggest cob should be used as seeds. It was believed that beginning about 2500 BC, perhaps earlier, the crop spread through much of the Americas. The region developed a trade network based on surplus and varieties of maize crops. Today, maize is widely cultivated worldwide, and a greater weight of maize is produced each year than any other grain. In 2014, total world production was 1.04 billion tonnes. Maize is the most widely grown grain crop throughout the Americas, with 361 million metric tons in the United States in 2014. Approximately 40% of the crop—130 million tons—is used for corn ethanol. And the development does not stop. Genetically modified maize comprised 85% of the maize planted in the United States in 2009.

When I took this photograph of the edge of a maize crop at Marondera, an agricultural field station in 1987, maize production in Zimbabwe was 2.3 million tonnes much of it exported. In 2022 the crop yielded 1.5 million tonnes and a further 500,000 tonnes had to be imported.

There is no way that those early farmers in South America could contemplate a billion tons of production from their small plant. Indeed, they would not recognize the modern varieties, but they knew what they had and exploited it for their benefit. Humans are genius at spotting an opportunity for personal gain. More food meant more people, but both only happened because people traded. They tapped each other's desire for betterment and created systems of trust.

Just a little thought, and it's amazing how our human commerce has not been cheated out of existence in a generation. Somehow, that system of trust worked. I'll give you this bag of maize seed in exchange for those spears, and I trust that you will not bring your mates and steal them back in the middle of the night. You worked hard to fashion those spears, and I toiled in my field; we respect each other's efforts. You value the seeds, and I value the spears equally. We have a fair trade. Back in those raw 'tooth and claw' days, trust to make deals was as essential to the success of our species as nimble brains, an eye for the unusual, and language. Indeed, we will never know what came first. What is clear is that trade persisted, and seeds were a key part. They were portable, could be stored for some time and were valuable. What more does an early entrepreneur need for success? There were cheats and bandits. We are still talking about humanity with vicious spite for every act of kindness. Still, trade worked for the majority because the speed of expansion, perhaps 6,000 years for maize to spread from southern Mexico to the whole of the Americas, is extraordinary. A kilometre a year in all directions. That could only happen with a trust-based system.

How good are humans? Not only do we see the unusual, we know when to exploit it and even how to get others to help us do it. No wonder success is in our DNA. We have experimented with it for thousands of years.

Hold on a minute. All this pontificating over our greatness is dandy, but it's not the whole story. If the successful dudes are away selling seeds for spears, who's looking after the maize crop in the ground? Their wives and children, obviously, and then, should they become wealthy and powerful, they might have a few slaves, too. No one ever said that the rise of *Homo sapiens* was moral or just, except perhaps white men and biblical scholars. However, the availability of labour has been key to agriculture for 6,000 years or more, and in some places still, fields are tilled by hand using sticks and mattocks. Power came from the people, so everyone was put to work. Not until around 3,500 BC did the first plough appear after the domestication of oxen and the draft power these animals offered. This invention was a boon to production. More challenging soils could be tilled and altered to receive a row of seeds. The fields could be extended, and more food was grown. More power came to the priests, and enough excess food was available to retain armies and build pyramids for the dead and ships to travel far and wide.

Modernity is fascinated by this development of ancient civilizations and the relics left behind: tombs and hieroglyphics, sarcophagi and golden ornaments. What often goes unmentioned is that this was possible only because of the complex agriculture. In the case of the Egyptians, a system we now call floodplain irrigation. The flooding of the Nile brought water and nutrients, allowing farmers to cultivate crops like wheat, barley, flax, and various vegetables. The Egyptians used a method called basin irrigation, where they created basins or trenches to trap and channel floodwaters to fields, ensuring even distribution of water and silt. This method facilitated the growth of crops in the nutrient-rich soil left behind after the floodwaters receded. Crop yields from wheat and barley were good enough to make them staple crops, providing the foundation for the Egyptian diet. Estimates suggest that wheat yields could range from 5 to 10 times the seed sown, showcasing the productivity of their agricultural techniques. Along with yields was an intricate system of management and distribution of agricultural resources, a taxation system where a portion of the harvest was collected as taxes, and storage of surplus crops in granaries to ensure food security in times of scarcity.

Although it will be debated endlessly, I think this is the start of the Anthropocene. As more humans discovered and prospered from agriculture, trade and civilizations came into being all over the world and with this success came more humans. There is nothing like reliable food that reduces infant mortality, allows people to live longer, and increases reproductive capacity. More humans need more agriculture and more fields of maize. And so began changes to the landscape from a natural mosaic of trees, shrubs, and grasses to a mosaic carved out for food production. This change was slow for a long time, barely noticeable in some parts of the world; likely, those 150 million were concentrated where growing food was easier, the soil was fertile, the climate favourable, and the crop choices resilient.

The rest of the planet was, as it always was, still merrily recovering from the previous mass extinction 66 million years ago, with the rare things minding their own business. Much of the world was inhabited by humans; even the deserts, mountains and the Arctic had innovative humans living in them. But most likely in small numbers and with a relatively small footprint, the same as any other predator and forager that might evolve in such habitats. There is conjecture about the demise of some megafauna. These large land mammals became dominant in the Pleistocene and appeared as mammal body sizes increased for some species after the dinosaurs left so much food on the table. On each continent, the arrival of people overlapped with these great beasts, and they coincidentally began to decline, only a handful of species managing to hang on into the present day. When humans arrived, many of these creatures were relatively abundant. Even in the late 18th century, there were 60 million bison in North America; not so fortunate were mammoths, woolly rhinoceros, giant tortoises, and many large flightless birds, including the famous dodo. Hunted to extinction in some cases, habitat destroyed in others, already on their way out due to climate shifts with humans a nudge too far.

Not so merrily, then? Human influence on some species was strong, even when we operated as foragers. This story grabs the headlines in many debates about our role in the loss of rare things. Indiscriminate humans gaining dominion. The savage is in need of reform. By this time, there were two broad groups of humans. Those operating as many had been as bands of mobile foragers across the landscape and the new style of humans who fed themselves from agriculture.

Just a glance at this image and you know that this technique was a lot easier than using a shovel. By Painter of the burial chamber of Sennedjem - The Yorck Project (2002) 10.000 Meisterwerke der Malerei (DVD-ROM), distributed by DIRECTMEDIA Publishing GmbH. ISBN: 3936122202., Public Domain, https://commons.wikimedia.org/w/index.php?curid=154346

So far, humans are tickling around the edges. They are involved in the extinction of a handful of species as they forage and cultivate across the landscape. But for the most part, the vegetation and its inhabitants are still intact. Organisms are doing their thing wherever they can. This continues through to historical times, where, for the most part, agriculture is still relatively confined to a handful of places. There are towns and even civilizations organised around them, but the overall ecological footprint of humanity is still relatively small. Even in the crusades of the Middle Ages, where great armies fought religious wars in the East, just a few thousand individuals were involved. Even the retinue supporting the knights and foot soldiers was relatively small by modern standards. However, the fact that those numbers of people could be mobile in that way, with food supplied to them as they went, was remarkable. Yet, still, the relative impact on the environment was small. By the time of the crusades, the Eastern Mediterranean campaigns between 1096 and 1271 to recover the Holy Land from Islamic rule, there were perhaps 350 million people on earth—most of them in India and China. And there were still parts of the world that humans had not reached. Madagascar, for example.

More than half the human population was still foraging rather than farming. This arrangement might have gone on like this: relatively benign footprints from humans increasing in number only slowly, mostly living short lives of toil and disease. Then, some clever people discovered coal. It might have taken a few decades, but such a resource changed everything. Particularly because not long after coal was exploited as an energy source, another enterprising individual found oil. Those two fuel sources, and later natural gas, were the sun's energy captured in the bodies of plants and animals and stored in rocks millions of years previously, a battery that we now call fossil fuels. That energy was now available to humans who started using it for activities beyond normal human capacities with their labour or a stick and a donkey. Humans now had copious exogenous energy.

So what happened next? What we did was we put that energy into the production of food. And, of course, with food production came more human beings. We achieved this numerical success by inventing and building engines through steam first, internal combustion, and later power stations generating electricity. Suddenly, there was unheard-of horsepower to till the soil, cut down trees, mill timber, and irrigate the land. Fossil fuel power, plus the technology to harness it, created an extraordinary shift in our use of the landscape. We now could move the earth around and ourselves. And so we did. In a hundred years or so, the landscape was radically altered from when fields were small and didn't extend into the difficult country to where fields became huge, operated by large machinery producing unimaginable yields. The result is today's billion tons of maize produced.

Global oil consumption is roughly 97 million barrels per day. If you laid 97 million oil drums end to end, you could go around the Earth's circumference approximately 2 times, with some distance remaining.

It wasn't just the fossil fuels that gave the energy to the land. Fuels also provide the energy to create fertilisers. During the process of developing agriculture, farmers realised fairly early on that there was a limit to how much could be grown in any one place. One solution was to move from one place to the next and allow the soil to replenish, what we now call shifting agriculture. The other option was to replenish soil nutrients with fertilisers. In modern agriculture, we are very precise about how much we add to the land to get the optimum yield from the crop we're trying to grow. The Dutch, for example, have, by the gram, information on the nitrogen, phosphorous, potassium, and trace elements that they need to grow food in their highly productive systems. They also benefit from having young soil high in clay minerals that exchange nutrients readily. On the other hand, Australian soils are generally old and have less capacity to exchange nutrients. But I digress. The point here is that the change to external power, that revolution in how agriculture could be handled, made a huge change to production. As humans increased, more land was converted to agriculture with this new energy source, and before long, we had radically changed the landscape.

We couldn't till all the land because sometimes the soil was poor or shallow or wouldn't yield crops, so humans husbanded livestock on open rangeland systems. Still, fossil fuels are used for efficient management of livestock. Grazing systems had been around for a long time, of course, by this point, and large parts of the earth would have been grazed by the domestic stock of one sort or another, but equally, they were grazed by native herbivores, too. So, the change in the management systems was more modest. The mixture of herbivores of different sizes and feeding preferences was replaced by livestock, typically of one species, occasionally two or three. In most places, these cattle, sheep, goats, and horses were not natives. Everywhere, this changed the plant species on the grasslands, even if the area was still classified as native pasture. But in those agricultural landscapes growing crops, the change was dramatic. Forests were cut down. The soil was tilled, and, in the end, a different ecosystem was created. One designed to channel as much biomass as possible into the yielding parts of the plants that humans wanted to grow.

Approximately 1,688 Panamax-sized bulk carrier ships would be needed to transport the entire 118 million tonnes of nitrogen fertilizer produced in 2021

Due to this appropriation of biomass, much has happened to the diversity of life, but before we get to any of the details, it is just worth reminding ourselves what the changes are.

At least 11 million km^2 of the world's land surface was converted to arable agriculture. In these areas, humans generated a monoculture only in the ground for part of the year instead of multi-layered vegetation with many plant species in each layer and the myriad of mostly tiny creatures that relied on them. The rest of the time the soil was bare, any weeds were ploughed in or blitzed with chemicals. Sometimes, a handful of enterprising farmers would plant a cover crop to protect the soil from wind, rain, heat, and desiccation. Any remaining native vegetation eventually became isolated and patchy. What started as crops in a sea of native forests became islands of trees in a sea of arable fields.

Another key feature goes with these changes. People owned the land, the livestock or both. Various tenure systems evolved worldwide, from commons to freeholds, but with ownership came demarcation. Areas that belonged were fenced or separated with stone walls. Animals were branded, and the concept of custodians and landlords emerged. Most people worked the land, while a few owned it. All this is familiar enough today to be obvious. The wealthy few and the populous workers. The psychology of the landlord and surf also had something to do with this story of the rare things.

It is also worth contemplating the speed of the changes. The first coal-fired steam engines were built in the early 1700s. The first petrol engines were in the late 1800s. Natural gas became a common fuel in the early 1900s. By the turn of the 20th century, when all this energy had fuelled the industrial revolution, there were 1.6 billion people on earth. Much of Europe, America, and China were already predominantly agricultural landscapes. Parts of South America and Africa were still undisturbed or dominated by shifting practices, but for the rare things that die was cast. And this is a blink in evolutionary time, far shorter than the time it took for the drivers of past mass extinctions to have their effect.

By the time Sir David embarked on the first Zoo Quest expedition in 1954, the conversion of net primary production for human needs was well established. This measure of human impact on rare things is more meaningful than dominion, land clearing, land use change, deforestation or any of the other emotive terms used to describe the process of agricultural intensification.

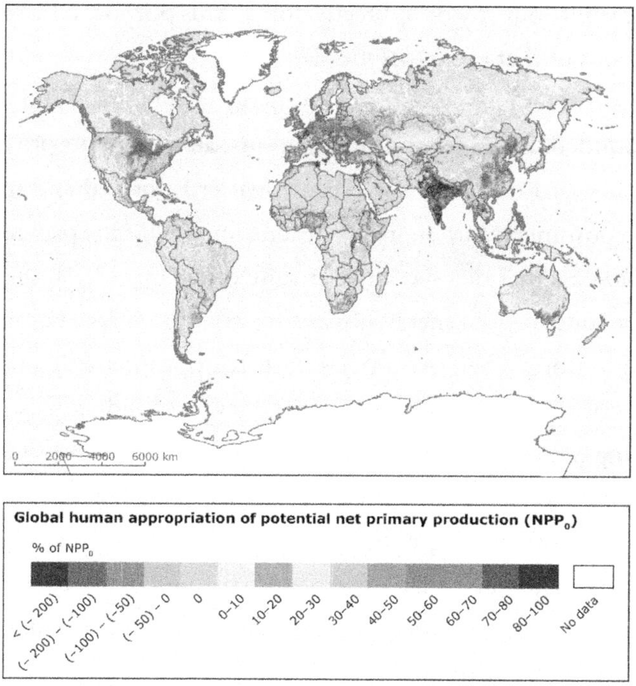

Map of human-appropriated net primary production (HANPP) as a percentage of potential net primary production (NPP). Source: European Environment Agency

Net primary production, or NPP, is gross primary production minus the rate of energy loss to metabolism and maintenance. In other words, it's the rate at which energy is stored as biomass by plants or other primary producers and made available to the consumers in the ecosystem. The difference between total photosynthesis and total respiration in an ecosystem is estimated by quantifying the new organic matter formed and retained by living plants in a given time. Then, humans come along and use the biomass. Human appropriation of net primary productivity (NPP) refers to the portion of the Earth's primary production (the amount of energy that plants capture through photosynthesis) that humans consume, directly or indirectly, for various purposes. Here are the proportions of that use across the planet.

What nature does is convert the sun's energy into biomass and then move it around among a myriad of organisms, all looking to progress their genes forward in time. New organic matter is formed and made available to consumers. Any biomass not consumed helps the plants stay alive; any biomass that dies has another suite of organisms waiting for it. This is the fundamental energy-dissipating structure we call nature. It is ecology in action. When one species starts to appropriate significant proportions of that biomass into the energy cycle of individuals, one obvious thing happens. There is less biomass for all the other species lined up to use it. Less biomass means less energy, and it's energy that determines the more-making ability of all organisms.

You might want to sit and take a deep breath before looking at the next graphic. It shows the distribution of mammal biomass on Earth in 2015, measured as tonnes of carbon.

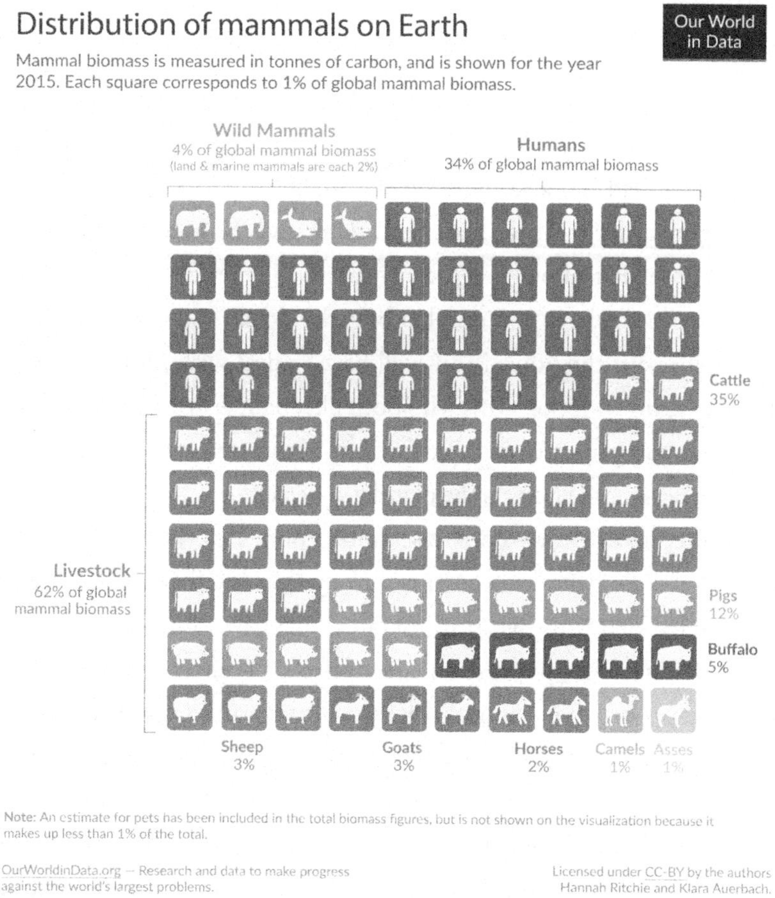

96% of the global mammal biomass is humans and livestock. Just 4% is spread among the other 6,495 mammal species, and half of that goes to whales and dolphins. It's no surprise that some of them are rare. I am not going to dwell on or explain this infographic just yet. Instead, I encourage you not to skim past it but to reflect a little on some of those percentages.

It's 8 am, and the family is off to school. Mum is also getting ready for work, and Dad has already escaped onto the 7:06 to central. Bugger, we forgot the kids' lunch boxes. It's Friday, and the lunch box items are depleted; the fruit bowl is especially sparse. There are three mandarins and three kids, phew. Kid number one, the eldest, takes two mandarins and rushes out the door. Mum will have to deal with him later. What should we do with the twins who both like mandarins? Mom engineers a delicate compromise to keep the peace. She promises to give both kids lollies after school and eats the mandarin herself.

Mother nature leaves it up to scramble competition where one of the twins loses. Pointing this out insults anyone's intelligence, but sometimes, it feels like those who care for rare things have forgotten this reality. Suppose the global population appropriates a quarter of potential net primary productivity and uses it to maintain and grow the number of humans on the planet. There is not enough biomass energy left to maintain all the other consumers of NPP, and some of them will not survive. It is blindingly obvious but too often forgotten. Meanwhile, the people who want to save the koala will also give to humanitarian causes.

It is worth reflecting on this psychology for a moment. We love babies, especially our babies and toddlers, doing cute things, which are almost as popular on social media as videos of cats. Only the truly disturbed would purposely harm a child. So, humans are the highest on the list of organisms to save. It may be that a qualification is required for human beings who look and think like the observer.

Helped inexorably by exogenous energy, humans have appropriated net primary production through agriculture and become locked into a cycle of growing food, producing more people requiring more food. Essentially, humans hog all the energy and the biomass. This story of people eating and drinking fossil fuels has several nuances. What is growing, where? What do people eat? Is what they eat nutritious, or even if it is a good thing to eat? Then, patterns of crop and livestock production techniques have changed to be more productive through intensification. Each of these things can take a whole research career to understand fully. Still, from the point of this narrative, we can accept that land use and the intensification of agriculture have been drivers of change in ecosystems. Change outside the evolutionary norms affects rare things, and in the process of change, the net primary production of the energy available to non-human organisms has declined massively.

Globally, around 13% of the total grain produced is fed to cattle. In the United States, approximately 40% of the corn crop and 70% of the soybean crop is used for livestock feed, with a significant portion going to cattle. In the European Union, around 62% of the total cereal production is used for animal feed.

But there is more.

Suppose an early farmer grew a crop, went to the marketplace with primitive versions of maize seeds, and tried to sell the seeds. If there are other farmers with seeds, selling them or getting a reasonable exchange is a struggle. If, on the other hand, there are few seeds available at that time at that place, then there is a good chance that the farmer can strike a tough bargain and gain a lot in exchange for the seeds. We take this process of market competition for granted. Modern markets are global—decisions about what commodities are and are not available are based on extraordinary knowledge. Prices are set in international markets, not the town square. This process of price setting, exchange value if you like, is a dominant part of the commercial model that humans adopted.

Suppose that there are ten farmers in the district, each producing maize. They all arrive at the market with their harvest at the same time of the year and try to sell or exchange those goods for the things they need. One farmer figures out that if he diverts the stream along the side of his fields when the dry season comes, he can put water into his maize plants, improving the crop. Once he's figured this out, he does it every year. He has more grain to sell at the market, so even if the price is low, he does better than the other farmers. In the years when the other farmers had no grain because the rains failed, he still had grain and essentially set his price. He doesn't realize or chooses not to notice that the water he diverted into his crops no longer flows down the stream. That may or may not have consequences depending on the river's size. But if it was a small stream and he diverted most of the water, farms and the land below the stream would be impacted during dry times.

That effect of channelling a resource into production that impacts outside of the production zone is called an externality. Humans have become extremely good at ensuring that as many costs as possible can be converted into externalities. In other words, costs that they don't have to pay, thereby increasing the net benefit of any trade. A modern example would be fertilizer use. Farmers will put fertilizers on the crop to maintain and improve yield. And bear no consequence for any fertiliser leakage from the soil system, polluting the streams and rivers.

Pollution is an externality. An outcome in the landscape, the environment, and the atmosphere is a consequence of business activity not born by the business itself. The cost is worn by the environment or other businesses in the market. When this cost is severe, such as in the 2000s when the production of chloride fluorocarbons was damaging the ozone layer with potentially disastrous effects for humans, companies were forced to deal with that externality and change the propellant use to non-damaging aerosols. In that particular example, the externality was reversed, and the ozone problem was resolved.

Most externalities are not so easily fixed, and they can be subtle. The effects are not easily attributed to particular business activities, and paying penalties for rule breaches is easier than resolving the pollution itself. As I write, an oil tanker has run aground in a high-quality marine habitat. The tanker had broken in two, making salvaging the vessel and recovering any remaining oil it carried extremely difficult. At any one time around the oceans, hundreds of oil tankers are progressing from one port to another. During the COVID crisis, when the use of oil declined, particularly in Europe and America, tankers were lined up behind each other, waiting to offload. They became floating storage tanks for the oil pumped out of the ground, just waiting to be used. These tankers transfer oil from source to market with a minimum of fuss. No one would suggest that that transport mechanism should stop as oil still powers most economies and moves people around. But now and again, there will be an accident. Historically, they have been dramatic and locally devastating for the environment, in which tankers run aground or oil wells spring a leak.

So, are those externalities? Yes, because they are a consequence of doing business—a risk. Nobody will know when, where or with what frequency such incidents will occur, but they still count as an externality to the business. And in some extreme cases, the companies involved end up paying reparations. But the thing is that there are no reparations to be had. Once the externality has impacted the environment, it is mostly irreversible. If it creates a change outside the norm, then there is the potential for the loss of individuals or even species. If it involves local extinctions, then by definition, it is irreversible. Remember that we don't know all the species in a habitat, only the iconic ones; there is usually unknown species loss. Even if the impact is mild, the ecosystem is unlikely to return to exactly what it was before. We'll come to this point again later.

Not only has the human population changed the land, but in the process of changing and in the market processes that follow, we have generated externalities where the true costs of those production systems are not born by the business but are born in part by the environment. Some conservation and green groups recognise this, and it explains why they are so against big business. Their anguish is understandable. Pollution events that are localised and acute, like an oil spill from a tanker, are unpleasant experiences for everyone—investors, owners, operators and, of course, anyone there doing the cleanup. Society loses, too, as nobody wants to see such a mess or environment damaged in that way. So, the Green lobby latches onto these situations and blames the corporations for their excesses and trying to keep as much of their business activity off that balance sheet as possible. They are big and ugly and only interested in maintaining profit. And you can be convinced with that argument. Some corporations are big enough to believe that they are untouchable, with dubious practices at best. Exploiting workers, exploiting the environment and not caring about very much at all other than the bottom line. There's a reason for this. A small number of people run large organisations, and a handful of individuals end up controlling the key decisions and choosing whether or not the company cleans up its mess. Again, we will come to that later.

Corporations result from the human choice to use market mechanisms to create the goods and services everyone enjoys. Even in those countries where those goods and services are in short supply and people are going hungry, people yearn for systems and opportunities to put food on the table for their families. The fundamentals of food and shelter are where the buck stops. Whether we like it or not, the current mechanism for putting food on people's tables is a capitalist commercial one that is now provided across the world along a six-continent supply chain that starts with intensive agriculture and the appropriation of NPP.

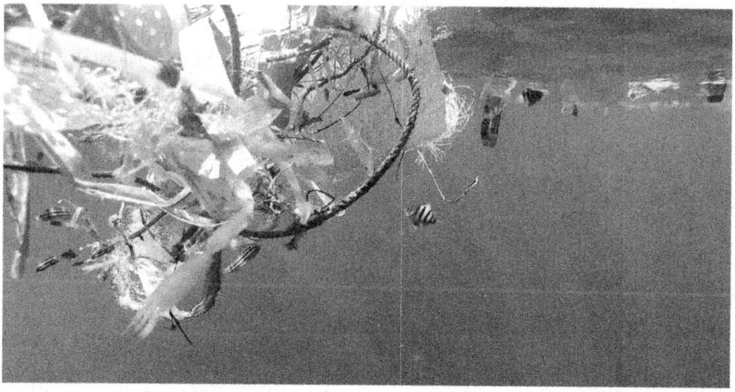

A 2015 study published in the journal Science estimated that between 4.8 and 12.7 million metric tons of plastic waste entered the ocean from coastal populations in 2010 alone. Source: Jambeck, J. R., Geyer, R., Wilcox, C., Siegler, T. R., Perryman, M., Andrady, A., ... & Law, K. L. (2015). Plastic waste inputs from land into the ocean. Science, 347(6223), 768-771.

There are communities where the old exchange style, barter, and simple living are lauded and available to some groups. But such an existence typically comes without health care, emergency management, and food from the supermarket in the drought. It also comes without the modern trappings that have made it possible for many people to persist. Exogenous energy use gave us more than extra food.

If a thousand people lined up side by side and were asked to step forward if they would be prepared to give up all the modern trappings and live a subsistence lifestyle, maybe a hundred would take a step. If you asked next if those hundred would be prepared to give up education for their kids, then maybe fifty would step forward. These keen folk are told that there's no education or healthcare under a subsistence system. And if there's a bushfire or a flood, you must look after the problem for yourself. This freaks a few more out; maybe twenty are brave enough to take another step. These are the true believers. Then, to these diehards, you say, oh and by the way, there's no mobile phones, electricity, or vehicles. You have to get by on your muscle power. No goods and services are available in the stores, and you have no food or money. The only help available is some draught power from animals you must rear and look after yourself. Otherwise, you're on your own. Maybe ten people will step forward.

That's the reality of the situation that we have created. Even if we all turn green overnight, living in the way people lived 12,000 years ago is not just unattractive; it is impossible. People will realize very quickly, as the 99% who at some point didn't step forward would show, that life without our economic system is not so great. Most of us accept the benefits of modern living and the externalities that brings. We may not agree with the costs and prefer a better system, perhaps one with a more equitable wealth distribution or fewer, less severe externalities. Some of us may want the world to be as pristine as it was 12,000 years ago. But we cannot have those things and still have our needs met. It is what it is.

The economic system exploits resources, changes the landscape, and creates externalities. These drivers change the composition of biology dramatically in most situations. There is no way such powerful forces can be ongoing for 10,000 years mildly and 200 years exponentially without an impact on the planet's biology. Sure enough, we have generated those changes, and it's not stopping. The rates of deforestation, the conversion of land, the use of fertilizer and the continuance of externalities are ongoing worldwide. Again, we could spend a lifetime researching the nuances of these drivers and their consequences, but the bottom line is that these changes are extensive and acute. There's nothing we can do about what has been and gone.

Many people, myself included, would like to transition to a different economic system. It would be marvellous to think that we could utilize what we know about money, how markets work, and how resources are used to devise a much more efficient and effective way of balancing the various values the environment delivers while keeping people fed and maintaining their well-being. This is especially important to the billions of people who would give up their lifestyles and live like you do, those in the euphemistically named global south. Not only that, but we're at eight billion and counting, so food production has to keep going to maintain food security across the globe; otherwise, we risk significant destabilization.

None of the alternative economic systems we've developed can get around that fundamental problem I mentioned earlier and passed swiftly. A handful of people make decisions, and a handful of people are the major beneficiaries. And it's not about money; these people are wealthy. The decisions they make are about power. They are psychological responses, not monetary ones. Any alternate system to counter the one we currently have begins with inequity. It has to recognize that somewhere along the line, the system has to cope with our fundamental genetic programming of more making, and that's some very deep psychology around control.

In the early days of agriculture, individual innovators learned how to grow food for their security and the security of their families. Staying put was an effective way of surviving what would have been a challenging environment. That didn't last very long. Even though the single-person family approach to agriculture survives to this day, what happened was that the method was appropriated. People could gain power by taking control of the resources they knew others were desperate to obtain. Such power has an intoxicating scent. Humans desire personal power above anything else. The invention of agriculture created an irresistible opportunity that, in various parts of the world, became commerce, a mechanism for people to gain power. And as a bonus for exchanging goods and services.

Commerce is always presented as a good thing with benefits everywhere. In modern times, commerce is essential as there is not enough land to keep people alive and grow vegetables in their backyards. Everyone is part of the marketplace that demands certain behaviours for efficiency and generates competition. It is fundamental now and worth understanding a little before considering the consequences of rare things.

In his excellent book, *But What Can I Do*, writer, mental health campaigner, and former Labor party spin doctor Alistair Campbell takes issue with the phrase 'it is what it is' as negative, a killer of initiative and because it implies that change is impossible. When you aim to get youngsters into politics, I would agree that all social change came about because someone initiated it. And it usually is someone, not a crowd, at least initially.

I use 'it is what it is' several times in this book about rare things. My use of the phrase is literal. The things we cannot change are inevitable. Extinction is inevitable. A human population of 8 billion souls requires 22 trillion kilocalories of energy from food daily. There will always be rare things, and not all will persist. My use of the phrase is the cold truth. It is pragmatic.

Here is my summary of the pragmatism we will need from Chapter 13 of my previous attempt to describe the environmental predicament humans have created, *Missing Something*, a book published in 2013. I want to end this chapter with a quote from it.

> Fossil fuels may be both a major source of greenhouse gases and the foundation of the global economy but they are a finite resource. At some time in the foreseeable future they will begin to run out. When they do, they will become scarce, and anything scarce but in demand can command a high price. This is good news for those peddling the scarce resource but bad news for the rest of us. The tension is obvious. If we are smart, we will engineer a transition to new energy sources that will ideally be clean, cheap, and ubiquitous. Sun, wind and waves are clean and ubiquitous, but at the moment are not cheap. Attempts at checking climate change by reducing greenhouse gas emissions put more pressure on the use of fossil fuels and narrow the economic gap with the cost of alternative sources. This is

a good strategic move irrespective of any potential effect on the composition of gases in the atmosphere.

An equally important strategic reason for a focus on emissions reduction is food and water security. In order to feed the expected 9 billion plus human inhabitants of the world in the coming decades, current global food production will need to double. If we keep to our dietary trends, then the proportion of animal protein in this food will also increase. Production of more food with a greater proportion of animal protein will need a lot of fresh water. We will have to take much greater care of our land, soil and water supply than we do at present. All that mining of soil carbon and nutrients will have to be replaced by farming systems that use the flows, not the stocks. We will be working to secure balance in the carbon cycle on individual fields, farms and catchments everywhere.

Chapter Five

The Willow Tit

Eric from Accrington is in the Horse and Hounds, thankful for his pint of Theakstons after a long day trip to the mere in search of a Bittern, *Botaurus stellaris*. He had seen this species a couple of times before, so the fact that he had a rough day, it rained, and the best he could do on anything rare was the call of a Willow Tit, *Poecile montanus*; only it was his mate Jim who heard that, and Jim is as deaf as a post.

"They're saying the Willow Tit will be extinct in a decade," Jim said through the froth caught in his beard.

"Really," Eric said, unsure what he wanted from his contemplative trance.

"Nothing they can do, they say. The reserves they put in didn't work. The numbers are still going down, losing their nest sites to Blue Tits, and their chicks get eaten by Spotted Woodpeckers. Did you hear that Fred, that old codger with a limp who lives around the corner from the Park chippy, claims that he saw one. The thing is tiny and lives in thick reeds; how will he see one? He struggles to get out of his car, poor bugger. Still, it would be a shame. Another one I wouldn't get."

"Who's they?" Eric said between sips.

"Aw, it was just the local paper. Read it when I visited Mom last week; she has trouble with her hemorrhoids again. Doctor reckons she might pass soon, but he has no idea how stubborn she is; she'll keep going forever just for the gossip."

"Did they have a source?"

"Sorry."

"How did they know the Willow Tit was going extinct?"

"I have no idea, Eric, what am I, some wacko scientist? How am I supposed to know where they got it from?"

I like Eric. He is a sceptic. He would not believe that his mother was dead until he saw her at the viewing. His first instinct is to question the source of any information, and over the years of living with this affliction, he has developed a nose for answering his doubts about opinions. His response is instinctual these days, but if asked to explain, he would put it down to the following few rules of thumb.

Rule number one: **Don't believe anything on Facebook**. This is a sensible start. Social media has no filters for the truth. Calling a random list of followers 'friends' says it all.

Rule number two is to **apply rule number one to all commercial media**. Fair enough. If it bleeds, it leads is hardly a motto for objectivity. No organization with a bottom line will do anything besides play on our emotions; they must keep our attention.

Rule number three is to **have a think and a gut feel.** This simple rule makes the most sense for anything, especially if you are prone to emotional responses. Well, except for social interactions involving the opposite sex. Eric is a little numb in the emotional department, so he tends to apply a lot of logic to his scepticism, which makes for tedious after-dinner conversation and may explain why all his girlfriends left him, but you cannot deny the rule; it is gold.

Rule number four is that **everyone is allowed an opinion, and I'm allowed to ignore them**. We could all use a bit of this enlightenment. Opinions are not facts, just feelings, or, as the definition goes, '*a view* *or judgment formed about something not necessarily based on fact or knowledge.'* Whilst there may be some truth in most opinions, there is too much of the other stuff to take them seriously.

Eric stops at rule number four. It's enough for him. In most everyday matters, it will suffice. He also makes the necessary exception to the rules for what, in his opinion, are erudite publications. *'Nature's Home'*, the RSPB quarterly magazine, tops the list, of course, as does anything to do with nature from the BBC. Thereafter, the broadsheets and The World Today on radio four. After many years, Eric has found no reason to doubt these sources. He is also a huge fan of Sir David and is disturbed that the great man is making pronouncements about the state of nature, but such a legend would not lie.

He looks up the Willow Tit on Google Scholar and finds reliable research evidence for perhaps 150 million pairs across continental Europe. So, it's not extinct then, just local losses.

The willow tit (Poecile montanus) is a passerine bird in the tit family, Paridae. It is a widespread and common resident breeder throughout temperate and subarctic Europe and across the Palearctic. So not at risk of extinction.

The point of bringing Eric back into the exposition of the rare things is that his scepticism is as rare as the birds he chases around the countryside. Most people do not follow his rules of thumb at all. Instead, they do the opposite and take almost everything as gospel. There are three reasons for this. Being a sceptic is hard work because thinking is involved. The sceptic's brain must stop what it was doing, gather information, filter the details for evidence, and then make a call. There is no simple reflex; some thought must be fueled and attention given for the second reason, a decision. A true sceptic must decide based on the evidence, even if the decision may go against instinct, what your mates think or legendary gut feeling. Making a decision has become one of the hardest actions in human existence.

When I was a kid, my father got his first car, a tiny Austin A40 Farina, first produced in 1958. That he had to wait so long for a set of wheels was a sign of the times and his chosen profession as a minister of religion. It was an illuminating childhood. One of the experiences I didn't block out of my memory was my father driving Austin into a multi-story car park, a rare building back then; it was nearly empty, and there were dozens of parking options. He drove up and down for five minutes before deciding on a spot near the stairwell—way too much thinking.

The third reason scepticism is uncommon is that it easily denies the feel-good we get when our worldview is reinforced. Scepticism denies us that high from cognitive dissonance, which is a real bummer. Abstaining from decisions to reinforce our worldview is a much easier route, so much easier. And the easy route makes evolutionary sense. Even on the savanna, the lioness walks down the tracks made by the tyres of the safari vehicles rather than tackle the grass wet with the morning dew—paths of least resistance. Hard work, decisions and the easy path make scepticism as rare as the Kagu, *Rhynochetos jubatus*, a near flightless bird from New Caledonia with no defence against introduced rats and cats.

Rather than embrace the curiosity and critical thinking of scepticism, it gets a bad name. Only the oblivious, like Eric, are overt with their questioning and demands for evidence. All the others keep their disbelief to themselves and ignore it long enough until it disappears.

What should happen is that scepticism—doubt as to the truth—should become 'no doubt as to the truth' because the observer has found sufficient evidence to make a decision. That is the process. Examine the facts, filter information into evidence, and decide whether it is true or false. It may end up as likely or unlikely, but that is fine. A likelihood is the best answer if the evidence is weak or equivocal. Either way, the doubt is tested and upheld or put aside.

Evidence for the start of the Anthropocene is unequivocal. The scale and level of environmental change in a single lifetime are similar in geological terms to a meteorite strike. When epochs are measured in tens of millions of years, a hundred years since the invention of the diesel tractor is only a slightly longer blink than a decade of devastation following the meteorite that ended the Cretaceous period 66 Mya. The local paper might be taking a punt on the back of an opinion or two on the Willow Tit, but that is fair enough, given the bigger picture. The chances of the earth coming out of the other end of humanity without a hit to biodiversity of at least three-quarters of species alive pre-agriculture are remote. Perspective is important in these matters because it may be possible to save a species here and there. In fact, and rather ironically, the super specialists with a narrow geographical distribution or habitat might be easier to save than the more general rarity. At least it is clear what habitat parcels must be protected and what conditions must be retained. Some SSSIs, sites of special scientific interest in the UK, are no bigger than a peanut to protect certain rarities. Fortunately, an orchid has no perception of a prison.

The doomsday scenario is hard to go past. The only conjecture is when it will manifest. As Sir David will attest, the process has begun. Research evidence shows species are going extinct at multiples of the estimated background rate. Certainly faster than evolution can replace them. So, net biodiversity loss continues. A rule of thumb rate is 0.3% per annum. That means a third of biodiversity will be lost in one hundred years and over half by the start of the 23rd century, by which time the rate will have slowed, and the last 30% has a chance of extra longevity. All this looks like the signature of a mass extinction event.

If anything is remarkable, it is that more species have not been lost already. That drive to survive and make more is so strong that species worldwide defy the odds and survive the human onslaught. But not for much longer. The extra push of introduced species and a nudge from the climate will be more than enough to see the end of many on Eric's list. He might even join a unique club started by the old sea dog on the sailing ship departing the island of Mauritius in the late 1600s, the last human to see a live Dodo.

When it comes to rare things, a high level of caution is needed for any of the claims made about their future. Most rare species are on an unavoidable trajectory to extinction. A few will hang on without much assistance, and a handful will be saved if great effort is expended on their behalf. Triage is the reality of the situation created by 12,000 years of agriculture and the explosion in human numbers and consumption that fossils fueled. We are what we are. It is time to start thinking about saving ourselves and paying attention to the species that will survive our onslaught, which is instructive. The rare things will not save us, but the species that can survive us might just.

Horror vacui is Latin for the better-known phrase 'nature abhors a vacuum'. This comes from Aristotle, who believed that nature contains no vacuums because the denser surrounding material continuum would immediately fill the incipient void. Ecologists get this idea, too. Nature hates it when a resource is left available to exploit. Put any organic material down on the ground, and it will decay. Once the energy of life that beats off entropy is lost, organic material returns to its original building blocks. This decomposition happens through the actions of microbes and fungi using that organic material for their own genetic goals. So when Aristotle says nature abhors a vacuum, he means that nature will come and fill in almost any space where a resource is available.

Yet, humans are incredulous when a pandemic appears, a virus simply using an available resource, when the apple rots in the fruit bowl or when shrubs invade the pasture. All around the world, people do everything possible to prevent nature from taking its course. And so we should, as there is no reason to say that preventable deaths and suffering from disease shouldn't happen. But we should also remember that when nature is presented with a significant resource, in this case, eight billion human beings, it will use that resource. Nature will always utilize anything based on carbon or anything that can be turned into carbon.

Rare things exist and are vulnerable for many reasons, but there will be no vacuum. There will be no loss of everything. Interpreting the loss of rare things as the end of everything is false. There are and always will be organisms in the landscape capable of utilizing the available resources.

Even when humans go extinct, which is inevitable, there will still be organisms left behind who will utilize whatever resources are still around. This has been true in every one of the major mass extinction events. From those deep losses came more diversity than was on the planet before the disturbance that caused the extinction, and the same will happen again after this current event. Even during the change and immediately after, when large amounts of diversity are lost, organisms are still present on the planet, and ecological processes will continue to function. If ecology can persist through ice ball earth, meteorite strikes, and massive changes to atmospheric composition, it can survive the consequences of a single species.

The organisms that will persist are already around. Some might be rare things that are only occasionally significant to how ecology functions because they are either highly localized or very few in number. The definition of rarity makes the rare species more likely to go extinct, but they might also make it. Rarity could even give some of them a better chance to survive in a pocket of opportunity.

For the most part, rare species don't have much to say about the broad ecological processes and their consequences in most landscapes. However, diversity is important. How much variety in a particular community influences how ecological processes of primary production, decomposition, and nutrient flow can be maintained, particularly when there are disturbances to those systems. For example, a grassland with many grass and forb species has a better chance of maintaining biomass production and nutrient exchange during drought conditions than a grassland with fewer species. The logic is that if you lose one species, there needs to be another one to take over its role seems to hold, at least in the broad conceptual sense. Ecologists have struggled to figure out exactly how true by conducting experiments with different numbers of species of plants and animals in communities, both artificially generated and manufactured in the field. The specifics are equivocal, but there is good evidence for the principle. The more diversity there is, the more likely the ecological processes in those systems will continue with less volatility and more resilience over time.

How does that work for the rare things? Well, the rare things do not necessarily contribute to that process. Unless they are rare until a disturbance occurs. As I mentioned earlier, in 2019, there were extensive bushfires in Australia, covering vast areas of forests in the east of the country. Some individual fires were over half a million hectares, and overall, 18.6 million hectares were burnt in the one fire season, an area the size of Syria and more than 30,000 km^2 more than England and Wales combined. The usual patchiness of wildfires was replaced by burning much larger areas and, as at Chris' place, exposing land normally immune from fire, such as wet gullies. In one of those wetter habitats, a species occurs that was rare enough to be listed as in danger of extinction, the Manning yellow solanum, *Solanum sulphureum*, a perennial shrub, with known populations confined to lowland rainforest and wet sclerophyll areas in the Taree—Wingham region, on the mid north coast of New South Wales. Before the fires, a few specimens were present, and the site was managed by Chris' team to assist in bush regeneration with the specific objective of maintaining the rare Solanum. In the subsequent months after the wildfire, they saw that the rare species proliferated. Indeed, it created an almost continuous stand in the ground layer across the entire site. This rare species was waiting for the right conditions to exploit the environment to its advantage. And while it was waiting, it was rare. The conditions came along, and suddenly, it is the most common species in the ground layer, providing structure and function to the habitat. It won't stay that way as other species return and out-compete it. In time, it will return to being rare.

There will always be common species. They may turn over in time, so a common species today may not be a common species 100 years or a thousand years from now, but there will be common species. These species will contribute significantly to transferring carbon and nutrients from the soil and the air into the biomass of that community. Their attributes and characteristics will contribute greatly to how carbon and nutrition are cycled around within that community. Nature, left to its own devices, will convert carbon dioxide into carbohydrates through photosynthesis fuelled by energy from the sun. Carbohydrates as biomass is a resource organisms will utilize thousands of different ways. If they fail to utilize it whilst the plant needs it, when it dies, or the leaf falls to the ground, another cascade of organisms will utilize that dead organic matter. Organic matter is a resource that will always be utilised, and the transference of organic matter in and around ecosystems is the process of life.

It is also the process that keeps humans alive and allows diversity to emerge over evolutionary time. It's the process that allows genetic material to become so divergent and creates extraordinary variety. Ecological processes are a given. They will always happen. Such surety is worth remembering when an emotional tug for a rare or cute species comes to mind. Carbon will always be pulled from the atmosphere. It will always be recycled, and there will always be organisms present to utilize the resource. If you think this doesn't work, then imagine a lava flow. A more inhospitable place for life is hard to imagine—scalding hot molten lava spewing across the landscape. Nothing can live in such a devilish place. But those lava flows cool, and eventually, bacteria and fungal spores will float in, as will plant seeds. The weathering process will begin to break down minerals in the lava. After some time, depending on the rainfall, temperature, moisture conditions, and the location of the lava flow, the soil will develop. Plants will begin to take root, and a succession of organisms will appear, and within a few thousand years or so, it becomes hard to see that there was even a lava flow there. The original molten rock is now solid, with a layer of soil above it and a layer of plants and animals above that. This is the way it has always worked, and it's the way it will always work.

All right, so if ecology doesn't stop, if the processing of nutrients and the movement of carbon from the atmosphere through the plants back to the atmosphere again is inevitable, why do we even need to worry about the rare things? If ecology is a given, surely there is no reason to be too concerned. Ah, but what about the moral obligation? Loss matters, especially if it is avoidable. The extraordinary philosophical challenge is to decide whether one sentient being has a moral obligation over another. Some would argue that maintaining diversity is required because diversity keeps the system stable. But that is our need, not one for nature. Ecology will always go on at the base level, even if its rates are erratic. And this happens with or without rare species. An ecosystem will function if a species is lost or declines in abundance. Another species could come along and do its job at a similar rate to the first species to maintain ecological processes even as conditions change. Or it might not, and the rates change, perhaps even collapsing into a cascade of species losses, and what was once a forest becomes a savanna, or a savanna becomes a desert grassland. Diversity might provide some resistance, but changes are inevitable, yet ecology continues. Eric is sure of it.

A farmer rearing livestock, let's say some merino sheep in a genetically controlled flock where the quality of each fleece has got to 16 microns and commands a high price at the market, wants to maintain his animals in the best possible condition. The biomass production in his fields matters, but so does feed quality because the grasses must provide nutrients and energy to the sheep. A diverse grassland with many species present is more likely to provide the nutrients the sheep need to grow a fleece that a human would call high quality. Diversity is important for production, which is a human use for that ecosystem. Similarly, if a farmer grows wheat, the native vegetation will be removed, and a single species will be grown. The original processes of nutrient exchange and biomass production will change, typically to an artificial transfer of added nutrients. This is faster and more efficient in producing the commodity, wheat grains, by channelling nutrients into a particular element of the original community. Such a system cannot persist for very long without significant input to prevent other organisms from invading that space and utilising the crop—from elephants to microbes, they will all be lining up for their slice of the biomass. So, the farmer ploughs the soil and applies herbicides and pesticides if ecology gets out of hand.

To grow the crop, the farmer creates a highly modified system only maintained with energy inputs. He adds nutrients because nutrients leave as the commodity is harvested and sent to market. Not only has the efficiency of nutrient cycling modified by the creation of monocultures, removing a large chunk of that diversity that recycles things efficiently, but a proportion of that nutrient is taken away at every harvest, be that a sheep fleece or a ton of grain. Those nutrients disappear from that soil, are consumed offsite, and are flushed down the toilet. But all the while, ecology has still happened.

Suppose that we stopped agriculture at a location. Ecological processes will continue, and diversity will return. Leave nature alone long enough, and it will fill the vacuum. Global figures suggest that roughly 40% of the original diversity of that location will return within a few decades. Nearly half of what was there before will return on its own without any assistance, and to the untrained eye, the vegetation will look similar to what was present originally. In most habitats on good soil, what returns are trees and shrubs, a herbaceous grass layer, and a wide variety of organisms that will be milling around using the resources. I say to the untrained eye because to know that species are different or absent requires careful observation and some experience in taxonomy, perhaps even a gentleman's demeanour.

Meat & Livestock Australia report there were 40.7 million breeding ewes in Australia in 2020, with 30.4 million (75%) of these being Merinos.

When humans modify ecological processes to feed and clothe humans, the disruption reduces diversity, often dramatically. I know this, Eric knows this, everyone knows it. However, ecology still happens, and organisms will return should the farm be abandoned or the production system be changed to favour nature. All the ecological processes continue, but not necessarily at the same rates or intensity that happened before agriculture. This is a problem for the rare species. Typically, they fill very narrow niches in environments and perform specific functions. And the opportunity for those functions may not be available in a secondary forest. They might also be energy-limited at the end of food chains, and plants in the secondary forest might be unable to fix enough energy.

That's, of course, assuming that specimens of that species are within spitting, flying or crawling distance of the recovering habitat.

A recap is needed before introducing one more consequence of agriculture that might end humans and any number of rare things.

People have moved around, spreading to all parts of the globe after the first wave of expansion out of Africa. And at remarkable speed. They gathered and hunted along the way and, thanks to language development, exchanged information from one group to the next. When some groups developed sedentary ways and gathered more information worth sharing or stealing, new information passed very quickly, hence the development of agriculture across the whole of the Americas in just a few centuries, a blink in the evolutionary process. In a very short period, modern humans had a system of surviving, the gathering model, and then of prospering, the agricultural model, that spread rapidly across the landscapes because both were efficient and provided exactly what people needed. It gave them the ability to have food and be food secure while having dominion and personal power. Although many people adopted stay-at-home agriculture, there was still a movement to trade, raid and generally show off newfound prosperity.

A couple of other things happened in this new system. The first was the domestication of various species that suited our purposes—livestock, of course, that provided food as protein and a walking larder. Livestock remains a critical protein source today, as do fish and seafood. But also draught animals, horses, oxen, and domestic dogs. *Canis familiaris* is now on every continent as both domestic working and companion dogs and as feral animals, wild in the landscape, living off whatever they can hunt.

Similarly, the domestication of cats was presumably to contain rodents in the new dwellings that humans constructed, which were as attractive as warm and dry places for mice and rats as for the people who built them. Cats have also escaped and live in viable feral populations across the globe. Humans transported the rats and mice, too, especially later in the story when humans ran out of local resources and began to explore the wilder habitats by land and sea. What better place for a rat than a wooden ship stacked high with grain for a long journey?

Crop varieties were traded. But along with them, perhaps mostly by accident, many other plant species come along for the ride. Some of these did well after being assisted with seed dispersal and became weeds—plants considered undesirable in a particular situation, 'a plant in the wrong place'. Not all introductions were accidental or a consequence of trade; many were on purpose. Some early European settlers to Australia wanted to remind themselves of the past glories of their homeland. They introduced Blackbirds, Robbins, and other familiar species into the landscape, including foxes. Descendents of these birds still exist and have spread far and wide. The foxes have become an expensive pest.

The introduction of species into areas that they previously did not inhabit is a problem for rare things. Successful new arrivals will appropriate resources, leaving less for the locals. So it is not just the human resource use, but the resource needs of their domesticated animals and the hangers-on that reduce what is available to native species. If the introduced species is a predator, then the results are obvious. Native species are vulnerable to a novel killer arriving in an environment without defence. In Australia, foxes and cats are the prime examples affecting and locally eradicating several species of medium to small-size marsupial species that, for thousands of years, had no significant predator of that size. Introductions are a compounding effect on the landscape changes brought about by agriculture. A novel predator in a natural environment is far less likely to impact local populations than a novel predator arriving in an altered landscape. The consequences of introduced species occupy thousands of researchers across the world. This is partly because of the conservation consequences on native species but also because introduced species are also a problem for production. The most difficult are species that we never see: viruses, bacteria, and fungal pathogens that afflict crops and livestock, along with their vectors and novel parasites.

From 1845, foxes were released into the wild in Australia and in 2022 there were roughly 1.7 million spread across 80% of the mainland and on 50 Australian islands. Research suggests they kill 368 million mammals each year from 114 species, or 40% of all land mammal species and half of all threatened mammal species on the continent.

Clearing for human land use appropriates net primary production to human needs and does more than change the landscape. It introduces novel species into that landscape and alters the likelihood of native species persistence. No wonder it is called the Anthropocene. The human enterprise is quite something.

And what was that consequence? It is the poor man's rainforest and home to a quarter of all the species on the planet. Eric tends to ignore it, and everyone calls it soil.

Chapter Six

Heavy lifting

Taxonomists have descriptions for more than 1,300 ant species in Australia. These insects are ubiquitous, abundant, and diverse on the island continent, with representatives of 100 genera, about a third of the global total for the taxonomic level up from species. As the majority of species are yet to be described and named, some experts expect the total number of species could be over 6,500. An ancient history, huge diversity and familiarity are not enough to get noticed by humans. Ant taxonomy is incomplete, and the exact number of species is unknown. This is partly because the survey effort is insufficient, and there are not enough specialists going around sampling and working through the complex taxonomic process necessary to describe a species and recognize it as different to the next sibling species. Another reason for the incomplete taxonomy is that some ant species are rare, either as super specialists or just uncommon, for one of the reasons already discussed. Imagine Eric's excitement if we still had half the bird species to discover.

Notwithstanding the incomplete taxonomy, Australia is famous for ants. They are everywhere, enjoying the dry conditions and the sandy soils. For such harsh habitats, they are a perfect organism with excellent heat tolerance, can survive with very little moisture and can use the weight of numbers to forage for almost any type of food. The social structure of the ant colonies makes ants operate more like a superorganism than a tiny insect crawling across your leg. They vary in size and shape and function enormously across the different habitat types, from the particularly nasty bull ants that both sting and bite to leave very painful welts between your toes through to the tiniest, the Big-headed Ant (*Pheidole* spp.) that has two worker casts with the minor cast as small as 0.75mm.

Ants have been around for a very long time, of course. The earliest fossil records are from the Jurassic, famously when dinosaurs were abundant, between 140 to 168 million years ago, when most plants were cone- and spore-bearing species like pines and ferns. They arrived just after Gondwana, the lush supercontinent, began to split apart. Africa and South America separated from the other half of Gondwana, opening the South Atlantic Ocean between them. Australia and Antarctica stayed together longer, not making their final split until about 45 million years ago. This ancient history explains why there are so many species of ants. All that is needed for diversity is the basic process of gene replication with occasional errors and drift, along with plenty of time and the process of natural selection of the better options among the mistakes.

Ants are also persistent. They could survive just about anything thanks to their body plan, social structure and ability to reproduce on mass. A fire ant queen can live for 7 years and produce as many as 1,000 eggs per day, which are tended by a workforce of up to 400,000 sterile female worker ants. Flying ants, or alates, as entomologists refer to them, are the reproductives of the colony, created by the queen and fed by the workers. When the colony is ready to expand, the winged male and female ants fly out of their colonies with a single purpose: to mate. This flight is coordinated among colonies into a swarm that helps lower the impact of predators via strength in numbers.

Along with efficient reproduction, ants can absorb shocks. Nests are often hidden away from disturbance and well protected. Ant soldier castes are not there for fun; they keep the colony from invasion by predators of various shapes and sizes. All this results in few specialist ant predators of any significant size or impact. Other ant species are usually the most common predators despite the cute echidna and bird species that take ants occasionally. There is another trick that ants have that explains why they don't need to distance socially. They have evolved the capacity to produce antibiotics from a gland on the side of the thorax. Antibiotics help prevent disease outbreaks within the colony. The meta-plural gland has been studied as a possible source of new antibiotics that could be synthesized for human use.

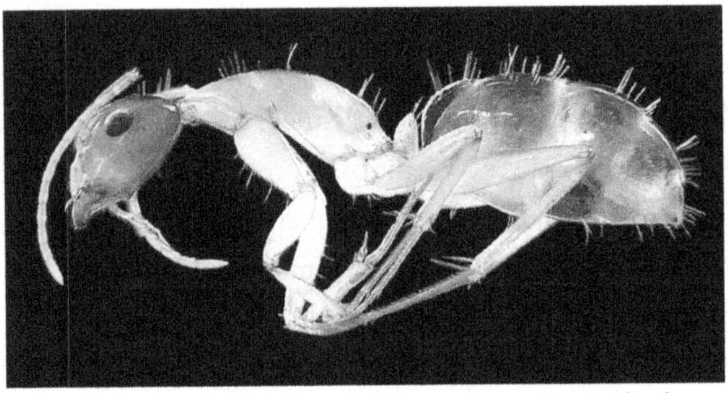

In the early 2000s I founded a company, Biotrack, that used smart collection and rapid identification of insects to generate a health index for natural and managed habitats. We used ants all the time because they were diverse and easy to sample.

Ants will stick around for a long time. Habitat destruction will send some species to extinction but not the group as a whole. Ants are flexible, adaptive, and robust and, like cockroaches, appear indestructible. Humans must hope this outcome applies to all the organisms that deliver or moderate key ecosystem functions.

Nature is dynamic, pulsed by temperature and rainfall, responsive to disturbance and, over the longer term, impacted by changes to climate. The interplay of cycles and disturbance creates the opportunity for natural selection to build the diversity we've been so fond of throughout this narrative.

That dynamic appears as a consequence of the basic ecology of natural and managed systems, the interactions between organisms and the environment that have been popping into this book at various points. In its simplest form, ecology is the transfer of nutrients and carbon, the recycling of those nutrients through the fixing of carbon by plants and the respiring of that carbon by animals. Not forgetting the decomposition of any material that doesn't make it into a herbivore. At its basic level, energy is captured to create order that defies entropy. This is not what the laws of physics say is supposed to happen, yet it does in a myriad of ways. Ecological patterns and functions are many and varied, and researchers have spent a long time figuring out how they work. And we'll continue to do so because the nuances fascinate us.

The point here is a simpler one. The benefits to humans from ecology, what we like to call ecosystem services, are delivered because of the structure and composition of each ecological system. Not just the physical structure but also the process structure that they display. A rainforest, for example, that has been rainforest for tens to hundreds of thousands of years, as most have, has a certain structure. Canopy trees, middle-layer trees, low-level trees, climbing plants, and epiphytes are complex three-dimensional structures with multitudes of organisms built on a relatively shallow active soil layer. And because of the temperatures and moisture, rainforests quickly recycle nutrients and carbon. The system is so efficient that not much carbon gets into deeper soil layers. Suppose the rainforest is cleared, perhaps in the Amazon, Costa Rica, or India. In that case, the soil will deliver nutrients for crops or livestock only for a short time, sometimes as few as half a dozen crop cycles, before inputs are needed, or a farmer must go to the next fresh patch of forest and clear it for his crops. This is because most of the rainforest's nutrients and carbon are in the biomass—the plants and animals. Take the trees away, and the rainforest will collapse.

A dynamic forest, with its complex three-dimensional structure, is iconic. We can see it in different rainforests all around the world. Each has a different species composition with a high proportion of rare things. However, the basic structure of the ecosystem is the same—rapid transfers through relatively nutrient- and carbon-poor soils into aboveground biomass. Only we don't see it. We tend to know about rainforests from images, perhaps the wonderful films on the Discovery Channel, because we have never been to one. We don't go to the wild; the closest most people get is to watch it from the couch. Human existence is now in a bubble. A barrier to the real world has emerged over the centuries of technological development, so most of us live separate from nature. This separation may have much to do with how we misunderstand ecology and how nature works.

A koala is cute and cuddly. Images of those doughy eyes pull on the heartstrings and make us feel good about ourselves. All that feelgood comes from pictures or an excursion to a wildlife park; hardly anyone sees a koala in the wild and so has no idea of the risks associated with that experience, snakes, leeches, mosquitoes, the occasional gum tree branch coming down in the wind.

A handful of people recognise the exploitation of nature and the moulding of landscapes into intensive production systems. Some of these folk say that such change is not morally acceptable, and they rail against it. But most people go to the supermarket or the marketplace to buy food. And, in the West, at least, have little idea of what it took to get that food onto their plates or what nature does for them because milk comes from the fridge. Our reliance on nature to process nutrients and carbon into products and services goes unnoticed. Services, from food to clean water and clean air to aesthetic value, are increasingly misunderstood. Especially how important maintenance of what scientists call ecosystem services will be for the future security of human beings on Earth. Instead, nature appears as documentaries on the bright and extraordinary presented by Sir David and his colleagues, who are trying to keep us aware of what nature has generated over the millennia.

Alright, so I can complain about a modern society that allows us to keep the dirt from our fingernails and learn little about how nature actually functions, even if such knowledge is foundational to human futures. But I should get back to the rare things. And now that ecology is in mind, some home truths.

Here is one—the **common species do most of the heavy lifting in ecosystems**. The common species achieve most of the primary production through photosynthesis, respiration and transfer of nutrients and energy. These common organisms are typically trees, shrubs, grasses, forbs, microbes and fungi, together with the consumers, the insects, including the ants, other invertebrate animals and a few vertebrates, also happen to be the most diverse groups of organisms. Once we leave species behind and start to think about nature as processes, the transfer of energy and nutrients, without which there would not be any rare species, then it is **the common types that matter**.

Size of the organism represents the number of species described in Australia for that taxonomic group

To illustrate this point, I would add a functional angle to this illustration modified from Professor Andy Beattie's excellent book, Australia's Living Wealth. The organism's size in the diagram represents the number of described species in that group within Australia. The microbes, fungi, crustaceans, and many smaller invertebrate organisms are diverse, as are the insects and the flowering plants. Indeed, insects and flowering plants comprise the bulk of non-microbe diversity in almost every system on earth. Functionally, these diverse groups also do the heavy lifting of capturing and moving energy through nature. The animals that make headlines and cause concern as rare things are often at the other end of the diagram. The birds, mammals and reptiles are the least diverse of the major life forms. It is hard to grasp but true.

The modest diversity of the iconic animals makes sense, given the late arrival of these groups to the evolutionary party. Insects are at least as old as the Devonian 412 mya, the first flowering plants known from around 140 mya from ancestors that separated from Gymnosperms 245 mya. The mammals begin from a single genetic line 210 mya, with placental mammals putting on evolutionary radiation immediately after the extinction of the dinosaurs at the end of the Cretaceous, about 66 million years ago. The insects and the flowering plants have had a lot longer to diversify and to speciate, and they have. In any one habitat, woodland, for example, a suite of species dominate the canopy. Trees, shrubs, grasses and forbes are diverse groups. And there's a very good chance that should a particular species be lost in that structural form; it will be replaced by another one from not too far away. Without human interference, the structure of communities would stay relatively stable over time. And even if certain species are lost, others will replace them. There is plenty of redundancy in the system. Not only are they common in delivering services, but plenty of species are available to fill the gap should a species be lost.

Why is that? Well, nature only happens because energy from the sun is fixed into biomass. If that fails, so does everything else. And we know nature abhors a vacuum, so if there is an opportunity to fix energy, evolution will supply an organism or six to achieve it. This is missed in the rarity story. Although rare species are at high risk of extinction, the species that deliver the key services are diverse or robust. What we're calling services are just nature doing what nature does—organisms moving organic material and nutrients around as they defy entropy and create more nature. It all begins with the organisms that capture energy from the sun using water to move nutrients around and generate carbohydrates for secondary consumers to respire. In other words, the basics are basic and will continue irrespective of changes to the species delivering them.

Human intervention to clear vegetation can neutralize processes whilst adding pesticides and herbicides into systems to maintain their simplicity for our ends. Still, after a few generations of letting those systems run fallow, they will return to something like the original. Nature will return to processing carbon at rates and configurations similar to those it did in the past. The premise that rare things are vital to our existence is untrue. Rare things are valuable to us for their uniqueness, but they are not a requirement for human existence.

This is a challenging concept to get across. The preservation and conservation of rare forms have value, as does the retention of wilderness. Still, that process is not critical to our survival now that the fossil fuel pulse has given us 8 billion and counting. Our survival depends on looking after the diversity and capabilities of the common species, the main groups of plants and animals that deliver the ecological processes we rely on.

Here is another home truth we arrived at at the end of the last chapter. Many of the species that do the heavy lifting are in the soil. And that's a place that, for many reasons, we don't seem to want to go.

Soil is much more than dirt, muddy boots and potatoes needing a wash. Soil is a complex material full of life. Just because it gets under our fingernails and is gritty, we think it is mostly mineral and lifeless. Not so; soil is a biological wonderland of organisms from amoeba to ants. And, of course, the matrix where roots grow in and around the smallest spaces to secure nutrients and water for the plant. Even the damp, cold soil under a winter wheat field contains biodiversity on par with tropical rainforests. But because we need electron microscopes to view most of the tiny critters and fancy chemistry to understand the variety of all the microbes, most of this life goes about its business unseen and unknown.

In a teaspoon of soil is dirt—sand, silt and clay—plus some dead plant material, dead animals, fecal matter and water. The amounts of each vary depending on the soil, for there are many types, honed from the parent material by the weather, but they are all present. Then, there is life, the organisms including 100+ million bacteria, 150+ micrograms of fungal hyphae, 10,000+ Protozoa, over 100 Nematodes, 150+ Arthropods (mites, collembola, insects) and perhaps a dozen other invertebrates (earthworms, slaters etc). Scale this teaspoon up to one hectare, and the mass of the bacteria alone can be the equivalent of 5 cows per hectare. For comparison, the recommended 'average' area of pasture for a single Jersey cow in the US starts at one hectare.

Then, there is the diversity of these organisms. The taxonomic groupings for numbers I quoted were very broad. If we drop down to the species level, the taxonomic resolution we are more familiar with for birds, mammals, and most plants and have defaulted to in this book so far, then the teaspoon of soil will typically contain **thousands of species**, many not yet described by taxonomists. Many are rare, just as rare and, in some cases, as endangered as the iconic species we care most about. They are still species but not as charismatic as snow leopards or koalas.

All this biology runs on carbon that comes primarily from the biological activity of plants fixing carbon dioxide from the atmosphere into plant tissue through photosynthesis and arriving in soil as dead leaves, twigs, dead roots but also as root exudates, compounds including sugars, organic acids, amino acids, enzymes, and other chemicals that plants secrete to communicate with soil microbes, alter the soil environment, and facilitate interactions with other organisms in the rhizosphere, the soil region influenced by root secretions. These days, we pay attention to atmospheric carbon, and we sometimes call this pulling of CO_2 into vegetation and soil via the action of plants 'biosequestration'. The carbon from biosequestration can be short-lived in the soil as plant tissue dies and is prone to decomposition aided by the activity of the many soil organisms. This is the normal, natural cycle of nutrients and carbon within the soil that maintains plant productivity. Without it, the plants would 'drown' in dead leaves and fine roots. Soil carbon also includes more stable carbon that can persist in the soil for millennia. This is the carbon from certain types of plant material, such as lignified wood in tree trunks and the inorganic charcoal and ash from wildfires. This so-called 'recalcitrant carbon' is an important long-term carbon pool in the global carbon cycle.

Organic carbon is the carbon in the soil that comes from biological activity—plant roots, root exudate, soil microbial biomass (bacteria and fungi), soil animals, dead organic material from plants (roots, leaves and twigs), and dead animals (mostly among those living in the soil). It is a strong indicator of soil health, where health means something important to humans. Healthy soil will absorb and hold moisture, be biologically active so that nutrients are available to plant roots, have a good structure to allow roots to penetrate, and provide a stable base for plants to grow toward the light. Together, these properties sustain predictable plant growth. Consistent proportions of inorganic and organic carbon are important because the inorganic carbon helps provide the physical structure and the moisture and nutrient exchange abilities, whilst the organic carbon fuels the biological activity. Both are critical to maintaining plant growth.

It is trite to say that soil is the foundation of agriculture, given that humans have cultivated soil to grow crops for 12,000 years. What is less obvious is that this cultivation tends to decrease the carbon in the soil. When natural vegetation is cleared and the soil ploughed for a crop, the carbon in the soil is exposed. The organic parts tend to dry out and oxidize; losing moisture means many organisms die out, too. Soil scientists believe that over time, there has been a net loss of carbon of at least 42 gigatons (154 gigatons CO_2e) from the soil due to global agricultural activity—nearly 5 times the total greenhouse gas emissions from human activity to 2008.

Around a third of the carbon in trees and shrubs is belowground in the root system

In most cases, lost carbon also means loss of structure and biological activity that makes nutrients scarce. Farmers learned to subsidize this loss by adding fertilizer. At first, this involved manure from the newly domesticated livestock and more recently, in the last 200 years of the 12,000 since agriculture began, as inorganic fertilizers manufactured on an industrial scale. Instead of healthy soil with biology cycling the nutrients, many modern farms are input-output systems where to produce a crop, the farmer must ensure that he adds sufficient fertilizer to the soil as an input.

Soil is the foundation of human societies, yet we mostly ignore it as a grubby black box. Eric and his birdwatchers no longer look down now that they wear Gortex boots. Advocates will bash heads with governments to encourage awareness of soil health and the importance of soil management to our long-term futures. But most of the time, governments and bureaucracies don't know where to put soil. Down the hall with the Agriculture Department, who are homogenizing the landscape for the best possible yield. Perhaps with the Department of Wildlife & Conservation staff chasing down the rarest mammal, or even the Department of Natural Resources, that nobody has heard of since the scandal with the permanent secretary.

Soil is the most important omission in our understanding of nature. Soil is where food production must reside for the bulk of global food security for as long as it takes humanity to figure out an alternative form of production to the tilled field and rangeland. But we don't seem to care. The UK has a 60 million pound-a-year budget to monitor water quality in the landscape and a $200,000 budget to monitor soils. That the soils are almost completely responsible for that clean water, acting as a filter for most of the water that flows through every catchment, seems to have missed people's attention. They'll measure the ultimate effects rather than the proximate cause. We have to turn this around to understand how the foundation of nature on land supports our needs and wants. Sure, lamentations and grief over the loss of species are real; we will be sad at that and, for some, unhappy at changes to the landscape. This loss is not felt by most. Average John and Janes, going about their normal business, need to learn that milk does not come from the fridge; it starts in the soil, and without functioning soils, the cows run dry.

In Australia, the dairy industry provided an interesting example of how little is understood about soil. For some time, milk yield was related to the amount of fertilizer applied to the pasture the dairy cows grazed. Adding more nitrogen resulted in more grass with good nutrition that produced more milk per cow. Understandably, dairy farmers were keen to add nitrogen fertilizer to the pasture because yield was a critical component of their business. Australia has three major food and grocery retailers with a combined market share of over 75% of the food retail industry. Two of them decided milk was a great loss leader and pushed the price down to as little as $1 a litre on the supermarket shelf, a price representing a few cents a litre at the farm gate. Milk producers on strict forward contracts could not cover costs, let alone make a profit. Naturally, some of them chose to break out of this market failure by doing what they could to increase yield. They added more nitrogen to their pasture. At least they could get away with volume, even if the prices were low. Only they didn't understand what was happening in their soils.

The many years of adding nitrogen to pasture had not only produced milk but also allowed the grass to develop strong root systems and increase the organic matter in the soil, in some areas to more than 20 times the background level for those soil types. In many respects, bulking up the soil carbon is a good thing, not least for carbon sequestration from a climate point of view but also because more carbon is usually beneficial to production over time. Soils with more carbon tend to hold onto moisture better, exchange nutrients more efficiently, and are biologically active, which in turn improves soil structure for plant growth. However, as with most biological processes, there is a limit, as there can be too much of a good thing. In those soils where the carbon levels had increased dramatically, the carbon connected with the nitrogen that was added and locked that nitrogen up into chemical complexes, which meant it was not available to plant roots. Farmers were adding nitrogen and seeing no benefit in the yield of grass or milk. Still desperate to get out of the market failure through more yield, many invested in wasted fertilizer additions until the industry peak bodies managed to get the message out to cut down on nitrogen.

Australia's 5,213 dairy farms support an overall dairy herd of 1.44 million cows. The typical Australian dairy herd has more than tripled in size in the past 3 decades to average 276 cows in 2019.

This example is a caricature. The plight of the Australian dairy industry and the failure of the market to protect it was highlighted in the press. Consumers were quite comfortable paying more for their milk, and milk prices have risen since the acute times and now reflect something closer to a respectable market price where producers can make a living. But the soil story was a salient one. It said several things about our understanding. First, we didn't have very much. Second, just pushing a particular solution to its extreme will not necessarily deliver the desired outcome—there are limits to ecological processes. Third, decisions in the supply chain can be many steps removed from real consequences to the producer. Fourth and most importantly, we know much less about production than we think, especially about ecological limits.

So why does a process of nutrient exchange falter like this? Adding so much nitrogen pushes the grassland well above anything that the natural systems have been exposed to over ecological, even evolutionary time. It is a shock to the system, just one of the many that are part of extinction events. Ecological shocks have several consequences. They tend to change the species composition, in this case, of microbes, fungi, and soil animals that moderate the nutrient exchange with the plant roots; in turn, this alters the grass and forb species in the pasture. Most of the pastures would already be altered by management, and even if there is diversity among the heavy lifters, capability has a limit where the process can close down. This much was obvious to soil scientists and even some agronomists who understand soils. Soil biologists and soil ecologists would all say that such a disruption to the system would generate unknown consequences. A few would have been able to predict that nitrogen would compound with the carbon and become essentially locked and unavailable to plants. Technical and scientific knowledge was sufficient to understand that problem. But it wasn't widely known or persuasive enough to prevent many landholders from investing in nitrogen, inputs which did them no good at all.

This is a critical message. It's more than the rare things because we need to understand how actions impact what nature provides us inherently and how much of that we want to manipulate. How much of that we want to improve, and how much we want to keep going. The only consistent way to do any of this is to understand nature. Not to focus on the loss of the rare but on the performance of the many. Those abundant species that do all of this work are the source of our food security. How do they do it? How do they maintain ecosystem services, and what are the limits to maintaining those processes in each circumstance? Answers to these questions are critical to human survival and social cohesion. It would be sad, but we can live without the koala. We cannot survive without healthy soil.

The United Nations is often quoted as a source of truth. One of the numbers that the UN Food and Agriculture Organisation has provided for some years is that 40% of global agricultural soils are degraded. It is hard to feel the magnitude of such a statement. What do they mean by degraded? Is it that the soil is performing poorly for ecosystem processes that we've been discussing? Maybe operating below capability for that soil type. The soil varies in the movement of nutrients and carbon storage. The different combinations of sand, silt, and clay, even the types of clay in the soil, influence the capacity to hold and facilitate biology. And each soil type depends on the weather and climate above it. Given the enormous range of soil capability globally, 40% represents the performance of those soils against their capability. In some way, they're not producing as much human benefit as they could from biological processes, and 40% is not a small number. It's almost every second field that is not delivering what it could.

There's another trend that puts the 40% in starker contrast. Global population growth, shifts in diet, and wealth creation worldwide are increasing the demand for calories and nutrients from food. The global food production system will need to grow at 2% per annum for the next 30 years to meet the expected demand. This net gain in production of 2% for 30 is equivalent to the first green revolution in the 1950s when agricultural machinery, fertilizer manufacture, and pesticide use transformed agriculture. Humanity will need the equivalent of that transformation all over again. Only the easy options for intensification have been made, which the FAO implies have degraded the soil in one way or another.

Few people understand this, and it would not make anyone jump up and down with joy. It's a massive challenge. Whilst there are reasons to be pessimistic, the upside exists, as with most human predicaments. It begins with understanding the challenge, realising the magnitude of the task, and the investment and social shifts required to make it right. For example, food will need to be more expensive. This is absorbed easily enough in mature economies, given that households spend around 12% of their disposable income on food. A doubling of food prices for them is neither here nor there. There will be complaints about belt-tightening but no real change to lifestyles. In other parts of the world or for the disadvantaged in mature economies, where the cost of food and the availability of income is such that food purchases might make up 40% or more of weekly income, doubling food prices would be catastrophic. These are significant, not simple or tiny challenges.

The concept of Lebensraum which translates to "living space" was central to Nazi ideology in WW2. The logic was that the German people needed to expand their territory to secure sufficient agricultural land and other resources to support a growing population and ensure their long-term survival and prosperity.

Soils are the foundation. Sure, exogenous energy from fossils turbocharged human technology and numbers, but the food that made it possible came from the soil. Looking after those soils will be our saving grace, at least in maintaining the food supply and social stability. There's nothing like hunger to galvanize a population into action. Remember that most wars are started over fundamental access to resources. And land is grabbed not just because it beats back the enemy. That land is valuable.

I feel guilty saying this because I live in a Western democracy with abundant food, income, and the ability to purchase the goods and services I need and some that I want from all around the world. Only that sense of privilege has to be tempered by the reality of the situation that affects so many others. Most people are far less fortunate and are very concerned about their food security from day to day and what their grandchildren will eat. This risk comes from the degradation of ecological processes delivered by nature, all of it, not just the rare things.

It is time to temper our obsession with the endangered and become more concerned about ecological processes delivering services. To worry about what the bulk of the species do, the species that do the heavy lifting. How does nature maintain those services and the structures on which they are founded? And to put all of our energies into it. The irony, of course, is that if ecosystems are healthy, then many rare species will survive. So we get much of what we currently desire as a bonus—a co-benefit, as the jargon puts it.

I would hate to see elephants extinct in the wild. That would be a tragedy full of sadness, grief, and shame. But I don't want to see the fruition of mutually assured destruction that our ignorance and indifference are taking us. More than the rare things, it is a call to understand soils and ecological systems, especially the ecology of food, how biology drives food production, and how we can bring more of that biology into play to make food production systems more reliable and resilient.

Organisms exist because they channel energy to foil the natural tendency for energy to dissipate and return the universe to its constituent elements. Life defies entropy. This remarkable, energetic system retains the organization that has existed since life started. The primary pathways—photosynthesis, respiration, and decomposition—have been present since carbon-based life evolved. The main organism groups doing all the hard work have been around for a long time, and the heavy lifting will continue. Whatever the disturbance, species will be present to deliver the bulk of the net primary production in both natural and managed habitats. The soils will be diverse enough to cope with the future. However, there is concern that change is making it difficult for the heavy lifters to do their job. It would be a problem if habitats were now wetter, drier, hotter, or more fire-prone than in the past, such that no tree species locally can persist under the new disturbance regime. When opportunities are created, no species exists close enough to take them up. That is a problem. Nature has created plenty of redundancy in the processes that matter, and the persistence of nature is inevitable. The persistence of particular species is not.

From a moral policy or outrage perspective, the real problem with extinction is not that species go extinct but that **a given species** goes extinct. This means that humans raise the level of importance of a particular species above all the others because it has a higher chance of extinction. Now, if we think about this logic, there really isn't any, at least not for human benefit and survival.

The challenge has changed. Not that specific species persist, but enough species persist to deliver ecological processes at the rates needed to support humanity. This is dominion again only it goes from being the problem to the solution. We need to shift from a manic obsession with species that are about to go extinct, who are, by definition, doing very little in the landscape other than assuaging our guilt, to enough species. How many species do we need to keep in the landscapes for those landscapes to persist as functioning units? This question sounds callous with a voice from the brown end of town, and to some extent, it is. Only it is a reality for whatever human value is the focus. To keep people from starving, natural and managed systems must continue producing at and beyond current levels for at least another hundred years. Long enough for the demographic transition to be completed. What happens in the next 30 years will determine whether that's possible and how likely our survival is for the next hundred and thousands.

Scholars have said this in the past. Thomas Malthus famously wrote in 1798 that increasing a nation's food production improved the populace's well-being. Still, the improvement was temporary because it led to population growth, restoring the original per capita production level. Malthus realised that humans tended to utilize abundance for population growth rather than maintaining a high standard of living, a view known as the "*Malthusian trap*". We avoided the trap because wealth and well-being have grown along with our numbers. In the time of Malthus, when the fastest form of transport was a horse, trains first appeared in the early 1800s, and the prospect of a Boeing 787 Dreamliner flying non-stop from the new Antipodean colonies to London was unthinkable. Even the poorest people today live better than the poorest in the 1700s, and the rich are rich beyond the comprehension of ancient kings.

Only did we? What is forgotten is that Malthus lived when there were one billion people on Earth. There are 8 billion today, and the number is increasing at 8,000 an hour. Think about it. He was right.

I was born in 1961. In the six decades to 2022, the global population has grown by approximately 4.87 billion people at an average annual growth rate of about 1.59%.

A human population of 8 billion is unimaginably different to the 10 million or so hunter-gatherers meandering around in small tribes 50,000 years ago. As Nate Hagens is fond of saying, humanity is a superorganism with a voracious appetite for food and energy that batters more rare species into submission daily. This truth is uncomfortable. One way to feel better is to save some iconic species, especially if we assume that saving them comes with co-benefits. There is an argument that if rare things are saved, this will reverse biodiversity loss and maintain as much diversity as possible from what is left on the planet. Keep rare things from extinction, and it will also benefit common species. This seems logical and has become powerful. It gives hope to folk who also want to see the iconic species saved. It points to our moral requirement to conserve, the motivation at the root of green organizations, and even as humans exploit nature, we might do it sustainably by saving the rare that holds an umbrella over the rest. We all share the world, so sharing and caring is an obligation. Phew, that does feel better—here is $10 for the 'Save the Koala' collection.

The reality is there is no obligation at all. Biology is not in any way moral. It does not have a higher self; it is just a process of gene replication and the evolution of various types. The way those genes come together to produce complex organisms where the organism is the vehicle for the projection of genes into the future.

Organisms that have developed extensive brainpower to cope with the complexities of the environment do have the ability to go beyond basic biology. Human beings, of course, think all the time. We have developed a culture and a sense of morality. But our basic biology does not need that moral code. Biology does not need moral judgment; that is, of our own making, a consequence of our complex brain that can contemplate its demise and give us more than was originally bargained. The moral argument is easily extended to the point that if threatened and endangered species are supported and maintained in the environment by removing the drivers impacting them, everything else is resolved as well. Silver bullet syndrome is an interesting human trait that also exists in the climate debate. The world is warming. Most people are convinced that humans have triggered that warming by releasing fossil fuels, with evidence enough for only a few crusted-on deniers not to accept this logic. Most people also believe that if we remove that activity and stop pollution from using fossil fuels, then the warming problem will go away. The rhetoric around the emission reduction solution is exactly what you must say to convince people they need to change activity. You cannot ask somebody to change activity without a result. There has to be gain from the pain.

The same logic applies to rare species. Focus on the rare things and assume that if they are cared for to prevent loss, the rest of the biodiversity will be fine. It should be clear by now that this assumption is false. While we assumed that saving rare species would be enough, bad things will happen to many common species, especially small critters.

Bugs are yucky, and some are even dangerous to humans. There are poisonous spiders, scorpions that sting, bull ants that sting and bite, and other species that transfer parasites or are disease vectors. Loss of bugs is a good thing, surely? No more termites eating through the rafters or spiders sending young girls into screaming fits. No more pesky insects in your ear, mosquitoes biting your ankles or cockroaches in the kitchen. Butterflies are lovely, but bedbugs from your Air BnB are not. On balance, most city dwellers would vote for an insect apocalypse.

Of course, a moment's thought and these instinctive responses appear foolish. Have you any idea what fish eat? Or what lizards, amphibians, and bats eat? Did you know that insectivorous birds consume 500 million tonnes of prey annually or that somewhere between $235 and $577 billion worth of annual global food production relies on pollination from honey bees, native bees, and flies? And this is not half of it. Tiny creatures control pests, recycle organic material from dung to corpses, tree trunks and leaves, keep the soil healthy, disperse seeds, and the myriad of species that decompose dead organic matter stop us from being neck-deep in even more ugly material. All of this happens through the existence and actions of insects and their invertebrate cousins that we call bugs.

Around the world, entomologists are studying insects. We should be very proud of these fine folk, for without their understanding, it would be harder to manage many diseases transmitted by insects, resolve many pathogens, figure out how to assist insects in pollinating crops and, most importantly, support insects and their invertebrate cousins to maintain soil fertility. Then there are insect people you know about. The pest control folk who make sure the fly spray kills the flies and not us or know how to eliminate termites from the rafters. Sounding a bit posh and, dare I say a little ivory tower, 'entomologist' usually refers to the researchers who gather the data and sift through it to find evidence for the good, bad and ugly on the insects that share our spaces. So we can listen to them with some confidence. Not only are they spending their days with bugs, but they are also the right kind of sceptic using numbers to find inferences. Lately, the number of insects observed by entomologists has been in decline. There are fewer insects around to be studied. Data from long-term trapping in 63 German nature reserves confirms what many have casually observed in many parts of the world. Insect numbers are decreasing, and not just by a little bit; they are plummeting.

> Hallmann C.A., Sorg M., Jongejans E., Siepel H., Hofland N., Schwan H., Stenmans W., Müller A., Sumser H., Hörren T., Goulson D., de Kroon H. (2017) More than 75% decline over 27 years in total flying insect biomass in protected areas. PLoS ONE, 12(10), eo185809

Three quarters in a generation. If the Dow Jones Index of share prices had collapsed by three quarters, the sky would have fallen. Imagine trying to survive on a quarter of the wages you earned when you started out. Heaven forbid if the defence budget from the 1990s was reduced by all those billions. How scared and vulnerable would we feel?

"Not a problem," says Eric, who has seen his fair share of insectivorous birds foraging through the undergrowth. "The crawling insects will simply fill the space left by the loss of the flying ones; that's what you ecology types tell us all the time."

Perhaps. Equally, a loss in numbers does not necessarily mean a loss of function. Pollination only needs one bee to transfer pollen from the stamen to the stigma. Fewer mosquitos have to be a good thing, and those beetle larvae can't be doing that much to the soil when we have fertilisers. It is always very easy to play the ostrich. Only they are remarkable and very dumb birds. When a dramatic and material observation that applies to many key ecological processes becomes known, we dismiss it at our peril. Suppose we ignore these numbers just because we like the idea of fewer midges at summer evening picnics without looking deeper to discover what is happening. In that case, we increase the risk to our already precarious food security.

Here is a quote from an article by Dave Goulson, Professor of Biology at the University of Sussex and the author of *Silent Earth: Averting the Insect Apocalypse*

> *It should thus be of deep concern to all of us that insects appear to be disappearing from the earth at alarming rates. In Germany, flying insects have declined by 76% in just 26 years. In the UK, our more common butterfly populations have fallen by 46% since 1976, and the rarer ones by 77%, despite great efforts to protect them by conservationists. Thirteen bee species in the UK have gone extinct and more will probably follow. In the US, the celebrated monarch butterfly, famed for its annual migration between Mexico and Canada, has declined by more than 80% since the 1980s.*

The monarch population west of the Rockies is down 99.9% in the last couple of decades and looks headed for extinction within a year or two.

Professor Dave Goulson

Insectageddon, as some have dubbed this disaster. Three quarters. One half. 99.9%. These are catastrophic proportions for numbers but also function. Imagine if you lost 75% of your salary or half your superannuation. Panic would ensue if this were a proportional loss of income or wealth, but this loss of bugs barely registers. The newsfeeds make it a filler somewhere down the list, with few people even realising it matters. The truth is what the loss of insects tells us about the environment. It means that the human impact is now acute. Land clearing, pollution, global warming and the channelling of net primary production into human food have crippled the hardiest creatures, organisms that have been around for 350 million years. Entomologists are sure all these factors have had an effect but are landing on insecticide use in agriculture as the biggest culprit.

And there will be consequences. Insects are not just food for birds and bats; they are integral to the ecological balance, the processes that sustain plant life, and the likelihood of human persistence. I kid you not. Remember that most of the food to supply the 22 trillion kilocalories a day required by the global population of humans is grown in soil. And that only happens because the soil biodiversity facilitates the exchange of nutrients into plant roots. Deplete the soil biology, and agriculture becomes inert.

There are plenty of things to do about trends in insect numbers. Start with the obvious pressures and reduce the use of insecticides and the leakage of fertilisers into waterways and native vegetation around fields. Land-clearing decisions must be more intelligent and considered. Clearing native vegetation to make the developer a dollar, no matter how 'crucial' development is to the economy, or a landholder says that food can be grown if the trees are cut down. We must know these claims are valid and make informed decisions on the value propositions.

Insects are critical to decomposition and nutrient cycling, and they are also a nitrogen and protein source. Instead of eradicating them, cultivating the promising species might be an idea. How about restoring habitat not just for the cute and cuddly critters but also for the invertebrates? The handy part of this is that it is much easier to regenerate insect populations than tigers or Leadbeater's possums.

Another radical suggestion that permeated this book is to shift a big chunk of conservation effort from the rare species to the common ones that do all the heavy lifting for ecosystem services because biodiversity is not a synonym for conservation. One practical way to do this is to create a matrix of vegetation that can support a diversity of insect species—functional patches rather than corridors for critters that never use them. Above all these essential practicalities is the need to raise awareness of how nature works and the critical role invertebrates play. This knowledge must reach everyone, especially politicians and technocrats. They should start by learning all about soil biodiversity.

An insect apocalypse is a harbinger. It tells us a lot about the rare things and more than the rare things. It is a symptom of the intensification of agriculture and the scale of human exploitation of nature. Nothing should trivialise this loss. When the bees go, we are in deep trouble. When the decomposers are compromised, the flow of nutrients to crops is too. We need the bugs.

The loss of insect populations is the reason this chapter began with ants. Not because ants are virtually indestructible and will still be around long after humans leave the stage. The reason for ants as an explainer is that they are both diverse, occupying different habitat types and niches and abundant everywhere. Particularly in Australia, where at least half the species are unknown, not yet described, and some not yet found. They exist in landscapes, and no human knows that they are there other than to estimate that there are species present that we have yet to describe.

So when we see a loss of insects, we're not seeing the loss of rare things we know about, the demise of populations, or the demise of species familiar to us. Many of these organism groups contain species that have yet to be described by science. In other words, diversity is lost at unknown rates. A consequence of numerical decline will be the extinction of the rare things before any human has catalogued their existence, as well as bringing on a tranche of newly rare.

The loss of the many is a very serious issue for human survival. Ah, but they're creepy crawlies, critters, and flying things we don't like. This instinctual response of 'they're just bugs' is naive. Invertebrates are far closer to the processes that deliver our survival than koalas and elephants. Every single gram of soil has thousands of individual organisms that steadily process carbon and nutrients. Lose those organisms and ecological processes slow, sometimes to a stop, impacting the efficiency of primary production, no matter how many tons of fertiliser we add. Plant communities become less able to deliver the commodities that we need.

A future with fewer insects is perilous.

It would be good to think that protection of the rare would deal with this problem of the many. For that to be true, the processes we would go through to protect the rare must also be processes that will protect the many. However, there is no 'one-for-one' relationship between those things. Disturbance causes rarity, habitat loss causes rarity, introduced species causes rarity, and natural processes cause rarity. Which of those will we fix for a particular iconic species we would like to save?

As I write, a Friarbird, one of 18 relatively large species of honeyeaters in the genus *Philemon,* is in the tree just above me, picking insects off the trunk. It is probably the widespread Noisy Friarbird *Philemon corniculatus* that is rare in my backyard, although I am not enough of a birder to be sure of my identification. Do I need the Friarbird? The problem is that we don't know the answer to that question. We don't know exactly how many species we need to perform the functions we're after. Perhaps the safe answer is yes as a precaution. It is logic often used to justify conservation.

There's nothing inherently wrong with wanting to save a koala or with wanting to see a cute panda cub born in the zoo. And there is nothing wrong with a sense of morality about our impact on the planet. But there's something very wrong about not understanding nature. The truth of nature and the truth of the scale of the impact of human activity on nature. We can save a few icons in order to assuage our guilt. But that is not enough. It's time to realize that it's more than the rare things. Begin to understand what that actually means and to do something about it. And in the meantime, there is the small matter of growing, processing and distributing food to 8 billion people.

The genus Philemon, commonly known as friarbirds or leatherheads, comprises a group of 18 comparatively sizable honeyeaters distributed across Australia, Papua New Guinea, the eastern regions of Indonesia, and New Caledonia. Their diet consists of a variety of food sources, including nectar, insects and other invertebrates, flowers, fruit, and seeds.

Chapter Seven

Farmers

Joe is a farmer, a grazier, to be precise. He runs merino sheep on land in the Cooma-Monaro region of NSW, Australia, a small contribution to that 30 million-strong national herd. Poor and erratic rains, an ongoing weed problem from unpalatable grasses that reduce available forage, and livestock constantly vulnerable to blight are an unholy interlinked trinity of challenges that have given Joe several bad years. Practical problems have been with the business for a while and Joe works long and hard to resolve them.

On the upside Joe's sheep graze paddocks that have been in the family for generations. He was born and bred on the property and can tell you how he has improved the pastures with fertilizer, clever grazing management and time and money on weed control. These actions have helped, but they have created another problem of debt. Borrowing from the bank to improve future yield has put tremendous strain on the finances, pushing the debt-to-equity ratio to its highest level in the property's history. Weak cash flow from the run of poor years has not helped. It is also worth mentioning that Joe is 62 years old, a year or two older than the average for farm owners in the district. While he is still physically fit and strong and never complains about his bad back, Joe has glanced more than once at retirement.

The average age of Australian farmers at 58 is six years above the national average for other occupations—the average age of livestock farmers is 60.

What will happen to this man, his business, and the land he manages over the next decade? No one will know for sure until it happens. However, we can create some scenarios and propositions and assign a likelihood of their occurrence. Under business as usual, the most likely outcome is more of the same—a struggle to realise enough output from the farm to service the debt. It would take several years of above-average rainfall for nature to push out the weed problem and generate above-average wool and lamb for the market. Redressing the debt imbalance would need an even longer run of good years. There is always a chance of good rain, even under the increasingly erratic, warmer and drier climate that most regional climate models predict. However, in the average and below-average years, the business would likely run into deeper financial trouble.

Knowing that good rains are needed is not much help to Joe. Not even modern forecasting can confidently predict excellent or lousy rainfall years, so he is left with little more than a punt on which side of average the rains will be.

How to manage his flock with this uncertainty is not a simple decision. No one wants to play risk games with their livelihood or the family legacy; only here, they are inseparable. Joe can provide for his family and retain his heritage if he keeps going and it rains. If the poor years continue, then sooner or later, the bank will step in and foreclose. The tension is palpable, straining the mind and body. Joe is nugget tough and about as durable. He continues, of course, ignoring the realities of likelihoods and financial instruments. Nothing worth fighting for was ever easy.

Rather than guess wildly, it is possible to apply some predictive science to Joe's challenge. We took his circumstances, including his farm's soil conditions, weather and typical grazing patterns, as parameters in a predictive soil model. Applying some backcasting using historical weather data and some forecasting for future weather, with and without climate change scenarios, it was possible to understand what has and will happen to Joe's soil and to the pasture growing on it. Here is what Joe faces on his paddocks as above-ground biomass, an estimate of grass production.

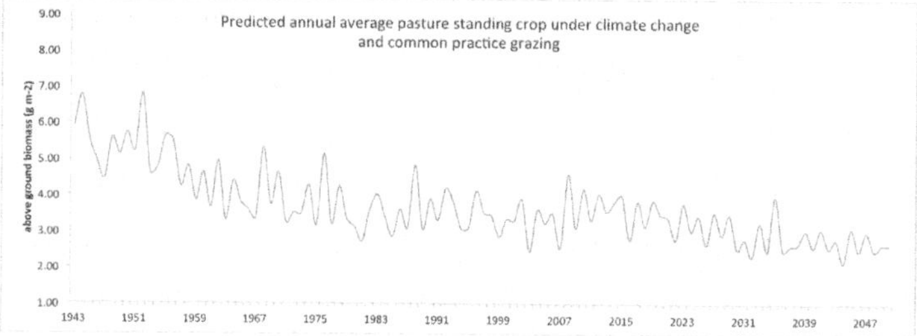

Joe must contend with a steady decline. The good years were excellent when Joe's grandfather was in his prime in the 1940s and 50s. Not every year, and you couldn't predict in advance what would come tumbling out of the sky, but there was always enough in the system to carry through the poor years. Today, Joe is under extreme stress from a run of bad years. He needs some good rains for wool production to service his growing debt and maybe even pay some down. He has become the first generation to see the bank own more of the farm than the family. But Joe is tough and tenacious. He has always accepted that farming is challenging in a land of drought and flooding rains. There is little point in complaining and even less in giving up.

Joe was born in 1955, and by the time he was old enough to notice, the bumper years were a thing of legend. There were good seasons, of course, only never quite the boon they once were. The poor years were harsher and seemed to last longer, and the soil was still dry in the following season. The decade of big hair, shoulder pads and Countdown was especially difficult.

When Joe's father died unexpectedly in 1989, Joe inherited the property and the responsibility. Joe had crisscrossed the paddocks, learning the ropes at his father's side for over 20 years. He knew what to do and that the land was only good for sheep, and nothing else would make sense unless deep pockets were handy to invest in land improvements. There were tweaks with fertiliser, pasture species, a little lime and rest periods for the most sensitive paddocks. Joe did them all for nearly three decades as time and money allowed.

Occasionally, there was a good year, even a run of above-average ones. There was a drought around the millennium that eventually broke. And all the time, there was no apparent pattern to the weather. Nothing Joe did seemed to turn things around or hint that good years were coming, but it didn't worsen. There was no detectable trend amongst the year-to-year noise. Here is what the soil model predicts for the yearly biomass production numbers since Joe's father passed. It confirms Joe's gut feel—variable and low but steady.

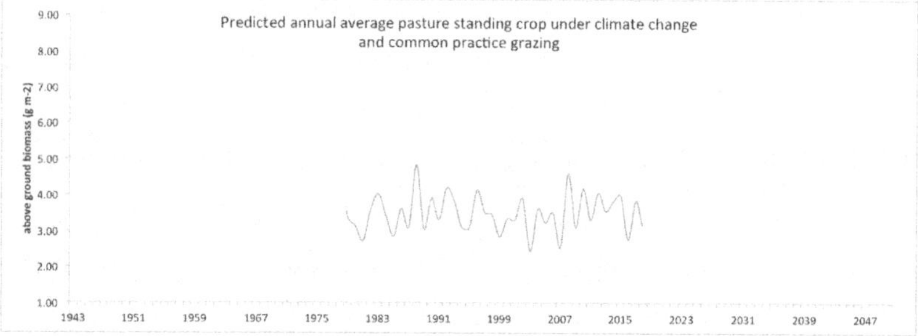

Treading water was survival, at least, and it toughened Joe's resolve. He dug deep and kept spraying the weeds.

Offering up a frown and a shake of the head, the government extension officer who paid Joe a visit clutching the modelling predictions pointed to the pattern over the next few decades. It was ugly. Average grass production on Joe's paddocks in the decade of the 2030s would be 30% less than today and less than half that in Grandad's heyday, assuming business-as-usual sheep production. The peak years will barely make it to Granddad's average.

Joe stared back at the officer. He does not believe this projection at all. His lifetime on these paddocks has been challenging, but the farm is still in the family and produces wool and lambs. There is no logical reason to believe that a turnaround in the weather can't see the good times return. Nature may be fickle, but farmers can be conquerors. Fair enough. Projections are abstractions, after all. Knowing how the grass will grow is in the lap of the gods, and no one has enough evidence to validate a future downturn.

Feeling his aching bones on a brisk winter morning, Joe is leaning against the side of his ute parked on a small hill in his easternmost paddock. Juliet, his daughter, is there too, staring across the property into the horizon with a slightly furrowed brow. Joe has spent a decade passing the skills, experience and sense of responsibility for the sheep and the land to Juliet. She is bright, keen and loves the land as much as her father. The bank demand for a loan repayment hangs in the air to spoil the view across the paddock. Rearing livestock in the Cooma-Monaro region of NSW, Australia, challenges the toughest, most tenacious of producers. It takes skill, experience, and attention to detail to sort out those who can and might not. Only there is a problem that no amount of capability can overcome. It contributes to the volatile grass growth, enables weeds, and affects the health and well-being of the whole property. A trend in Joe's paddocks affects the grass, and it is the biggest problem that Joe and his daughter must overcome. The soil model predicts that, over the years, Joe's family property has been steadily losing soil organic carbon.

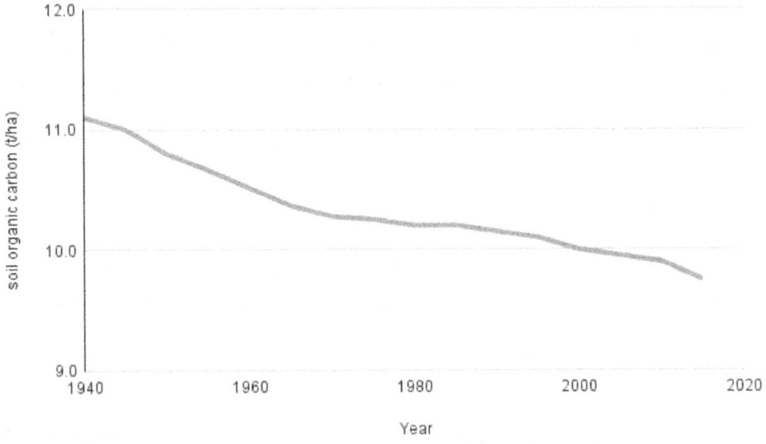

Trend in soil carbon on Joe's farm since his grandfather's time

Since Joe took over from his father in 1989, each hectare of land has lost 0.5 tonnes of vital organic material. Roughly 4% of the mid-80s level has gone. The steady outflow of carbon runs counter to the yearly bounce of the biomass to show a trend that a school kid could detect. At 9.6 tC ha^{-1}, soil carbon levels are 13% less than those in grandfather's time of bumper years, a loss consistent with what the boffins know about the complex interactions between soil, grass, grazing and the weather.

Dirt becomes soil when you add some biology. Life makes it possible for crops to grow and livestock to mature. Soil is the medium that holds the plants upright and provides them with nutrients, but the soil organisms make it all happen, and this is a vastly diverse community that runs on carbon. Soils get their carbon, also called soil organic matter or soil organic carbon, primarily from plants, as exudates from plant roots, root tissue turnover, and deposition of aboveground plant residues. This soil organic carbon is the glue, fuel, architecture, and foundation for complex biology that moves energy and nutrients around so plants can grow. How much of this organic material depends on the ecosystem, which in turn affects the biomass potential.

The synergy between soil organic carbon and the soil organisms that live on and from it is critical for plant growth. It is also essential for the breakdown of dead organic matter. If decomposition stops, then the whole production system collapses. Carbon also makes the soil more of a sponge to accept and retain water and a matrix to store and exchange plant nutrients. But you don't need the science. All country folk know that the dark, organic smell wafting off soil rubbed between the fingers is good.

Joe is baffled by this downward trend in soil carbon on his farm. He readily admits that his soil needs carbon, and no self-respecting farmer would want to lose any. Inordinate scepticism for anything in a white coat kicks in hard. Joe knows his soil feels the same, looks the same, and, as it has always done, varies from paddock to paddock and patch to patch. Sheepishly, the boffins admit that this trend is hard to measure. They say it would take 600 carefully placed soil samples taken in the 1940s and again today to verify the 1.5tC ha^{-1} loss over 70 years. A model is the only way to see the trend.

Joe sees a line on a graph that the boffins can't verify and calls bullshit. It's a trend, maybe. And so what? A ton of carbon in ten cannot mean anything too serious, and it is not as if the soil is blowing across the ditch to New Zealand. Juliet nods in agreement. Then, in the smart way that some brains process information, she asks the critical question, "Mr Boffin, can we reverse the trend?"

Well...

A month later, a new land cruiser is at Joe's farm gate. Three burly blokes in inappropriate city attire hover just outside the property boundary, a consequence of too many missed loan repayments. It is a wholly unpleasant scene. It should not have come to this. Joe does not deserve such a fate. An honest, tenacious and tough man, the head of an equally dedicated family that for generations has only thought of doing the right thing was brought down by a debt burden. The bumper years that would have turned it all around did not come. Family tragedy and sadness accompany the sorry deed.

Angry, bitter and overwhelmed by sorrow, Joe hands over the keys to the farm to the bailiffs. He watches the soft furnishings and other loose fittings of his life leave in the removalist's van that takes away what does not matter. Even the kelpies seem to struggle into his ute, as they somehow know.

The dust settles, and the farm falls silent.

The bank puts the property up in a fire sale, playing out the logic of its secured loan system. Only no one wants to buy it. Not even at a discounted price. A cash squeeze is one of the many consequences of the latest global financial crisis. There is simply no money available to borrow for such purchases, and cashed-up buyers are holding on to what they have now that cash deposits are, once again, positive. Foreigners are prohibited under rapidly enacted ownership laws to shore up the protectionism that always comes into play in a crisis.

Fencline contrasts are notoriously fickle but in this case the field on the left has been grazed continuously whilst the one on the right has been one year without livestock.

Juliet is waiting tables in town and couch surfing after she exceeds her brother's hospitality and patience. Being one of the few youngsters to keep his job at the bank has made him hard to live with anyway.

"Why not tell them my idea," she pleads.

"Because it's dumb, I would look like an idiot. You have no idea how easy it is to get fired."

"But it helps them out of a hole. They won't sell the place; some income is better than no income, even for a bank."

"What income? You can't run that place any better than Dad did."

"You reckon?" Juliette fired, then sighed. "Look, I don't suppose I could. It's just I know that the rest will bring back the good years. I know it."

It takes guts to grow food for others. At the end of June 2019, 89,400 agricultural businesses managed 384 million hectares of agricultural land across Australia. The average farmer was 58 years old, and three-quarters were male and had an average of close to four decades of experience in farming. The median age (the age at which half the population is older and half is younger) of the Australian population was 37 years. It's the old blokes like Joe that manage the land. This is not necessarily a bad thing. All that experience matters when the climate changes and the soil is straining to offer any help; all the tricks and skills learnt over the years are an advantage to maintaining the yields of crops and livestock. However, a generational change is about to happen. In ten years, it is unlikely that the average age will be 68; a large cohort is about to retire.

Farmers work hard and very long hours. They are up with the kookaburras and back in the dark. Half of all farmers work more than 49 hours per week—just 17% of the rest of the Australian workforce works that kind of shift despite being one of the longest-working people in the OECD countries.

Out on the paddock, it's hot in summer, freezing on a July morning, and either dry as three-day-old bread or an ankle-deep flood. Farmers absorb risk by the nature of their business—the personal risk of a manual job involving machinery and animals but, more critically, the financial risk from mostly small businesses. Over a third of Australian farm businesses turn over less than $50,000; many would barely generate the minimum wage. No matter. Most farmers would never give up their lifestyle. They enjoy spending their lives outdoors producing for themselves and their families. Growing food and fibre consumed and used by many is satisfying. The opportunity to make decent money in a good year is a bonus.

This life is unknown to most city people. It is hard to imagine. Perhaps we think it is marvellous that there are custodians of the land looking after it on our behalf. Very rarely do we think about it in terms of the cost to rural people. How resilient they need to be, and what growing all our food takes out of them and their families.

A while ago, while I was consulting with the Social Policy Research Centre at the University of NSW, Australia, we looked at suicide in rural communities, particularly among landholders. Rural suicide rates in Australia are significantly higher than those in urban areas. Researchers have identified geographic isolation, limited access to mental health services, socioeconomic disadvantage, and cultural stigma contributing to this disparity. Even if you want to, gaining regular access to a psychologist is hard if you must drive for two hours to visit one.

Suicide rates among Australian farmers are notably higher than those in the general population. Again, research suggests that the reasons include financial pressures from market fluctuations, droughts, or other environmental factors that impact crop yields and livestock production. As we have seen, farming is inherently unpredictable. The work is hard, with long hours and physical strain, and can lead to high levels of stress and mental health challenges. Often, the farmer has to do it alone. Physical isolation easily becomes social isolation, not made any easier by cultural values that reinforce the farmer as self-reliant and stoic, and may also discourage farmers from seeking assistance for mental health concerns, leading to underreporting and inadequate support for those struggling with suicidal thoughts. Mental health literacy, increasing access to mental health services in rural areas, providing financial assistance and resources during times of crisis, and fostering supportive networks within farming communities have all been suggested as interventions to reduce suicide rates among Australian farmers and promote mental well-being in rural areas.

Our conclusion from this work was that the sociology and the human psychology of food production are as critical to the process as the physical production of plants or animals. The cultural, social, and economic aspects of food production are what make the difference and where the decision-making happens. For example, suppose a landholder is offered a novel innovation, such as a technological solution to improve irrigation or crop growth through a slow-release fertilizer. In that case, he will not take that opportunity unless he is convinced it would reduce his risk. Any decision will be influenced by what he knows about his system, how much risk he already carries and whether or not he trusts the person who fronted at his farm gate with the innovation. In other words, it's a human choice as much as an obvious technical fix.

This is very important. We know that our farming systems and landscapes are designed by human psychology as much as by the logic of production or the maintenance of ecological processes. They come together based on what will make a dollar rather than what will be most efficient in the system. It's a challenge even to understand, let alone realise the consequences.

The rhetoric around rare species and the conservation of rare elements in the landscape sits uncomfortably in this tale of agricultural production. Joe's circumstances and response are typical. He is doing what he can with the skills and tools available, making decisions in his social and economic context as best he can. Farmers have done this ever since there was farming. Conservation-related matters will be a long way down the list of risks and concerns for farmers like Joe, even for those who understand and are sympathetic to conservation. Add to this the common assumption that rare things are rare because of how we've managed landscapes, and conflict of opinion is inevitable. It is easy to make the farmer the guilty party when you believe that habitat destruction has made the extinction of the koala inevitable.

But there is more than the rare things here. The conflict comes from the need of some to save rare things, but Joe's production system relies on species, too. Without the common, and occasionally some rare species, there is no biomass and no soil carbon. Species are critical to Joe's production system. Suppose conservation is to succeed in keeping the rare species. In that case, production systems must maintain their productivity and increase productivity and efficiency over time to maintain supply and meet the increasing demand. And the only way they do that is to lean into the ecology, to maintain and benefit from as many species as possible, especially in the soil.

Years pass, and the bank fails to sell Joe's farm. It lies idle. The water points dry out, and the paddocks see only wildlife and a few feral goats. Here is what the soil model predicts will happen to the soil carbon levels without the livestock.

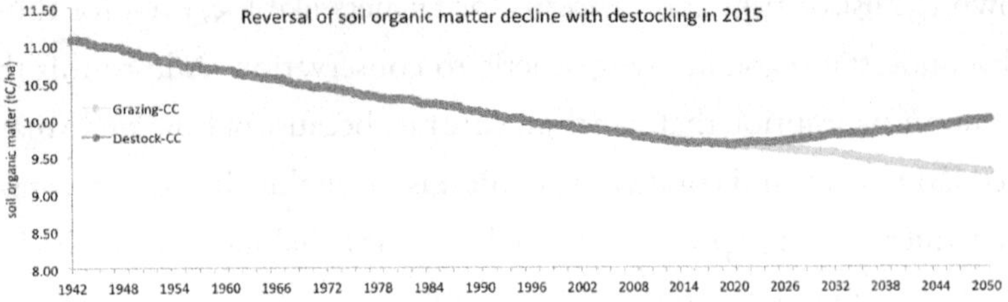

Reversal of soil organic matter decline with destocking in 2015
soil organic matter (tC/ha)
11.50
11.00
10.50
10.00
9.50
9.00
8.50
8.00
Grazing-CC
Destock-CC
1942 1948 1954 1960 1966 1972 1978 1984 1990 1996 2002 2008 2014 2020 2026 2032 2038 2044 2050

Chapter Eight

A Tinker's Cuss

Why do so many global cities have terrible air pollution, forcing residents to present at hospitals with respiratory ailments? Or that worldwide, 780 million people do not have access to an improved water source, and an estimated 2.5 billion lack access to reliable sanitation. A third of the people in the world need more water than they currently have. A third of the global population is suffering from some form of malnourishment, including 800 million people who go to bed hungry and 1.4 billion people who are overweight or obese. Presumably, this is unacceptable as a way of managing the environment for our collective well-being if at least one in three do not have clean air, water or the right food.

Then there are the local pollution events, with a terrible catalogue of oil spills, meltdowns of nuclear reactors, an ocean full of plastic waste, groundwater contamination, and asbestos in old roofs. There is chronic pollution from insecticides and agricultural runoff, and the biggest current pollutant is the emission of greenhouse gases combined with the conversion of land for agriculture, sending the CO_2 concentration in the atmosphere on an exponential trajectory. Not satisfied with making a mess of the land and water, in orbit, there are now 23,000 known items of space junk larger than 4 inches and maybe 500,000 smaller particles of debris—more than enough to protect us from the Klingons unless their shields are up.

Humanity seems to lack any understanding of what happens when we exploit nature. The current mindset that allows all these externalities is to convert natural capital into transferable currencies: commodities and cash. This is not to say that the alternative is to keep all rainforests pristine or not clear land for agriculture because we can't feed a global population of eight billion without food production. The point is that how natural resources are used is far more important than an argument about rare species. The decisions must be about nature itself, how to work with the natural capital, how to manage it, and how much of it can be used up to pursue a short-term gain before there is not enough left to support the people. I might sound like a neoliberal here, but there is more than profit in this logic. If people go without for too long or their living conditions are dire, they move to improve things. And if they can't move, they organise. A world of want is highly volatile.

Resource use is innate to biology. If there is a resource to use, organisms will exploit it. If the grass is on the plains, bison will eat that grass and make more bison. An estimated 60 million of them were in North American prairies before 1800 and the arrival of rifles. Bison follow instincts and will continue to feed, avoid predators, and reproduce as fast as the resources and predators allow. Humans exploit resources, but we could restrain from rampant resource use, particularly reproduction, to slow down and manage our population size. But we don't. Serious conversation about population control is taboo. Instead, everyone has the right to freedom and reproduction; end of conversation.

Fair enough. Rights are important and have been trodden on so badly for so many that their protection is both understandable and desirable. The challenge for the rare things is that they compete with people for resources such as space, food, and shelter. Most of them are at the wrong end of the global ecological food chain that has been appropriated so dramatically by humans—96% of mammal biomass on earth is now humans and their livestock, and 75% of the biomass of birds comes from domesticated poultry—and many of the rest are impacted by the externalities. This is the story the conversation movement tells us every day. What they don't say is why. They don't mention the 8 billion growing in number at a rate of 8,000 per hour and their rights.

Bison herds ranged across the Great Plains, from Mexico to Canada, and from the Rocky Mountains to the eastern woodlands. Estimates are that there were at least 30 million of them, most likely 60 million. Wildebeest populations in the Serengeti before significant human impacts numbered 2.5 million. By the late 19th century, bison were nearly extinct due to overhunting, with populations reduced to just a few hundred individuals.

In 1929, American demographer Warren Thompson described the demographic transition model of human population change over time. The model starts in the pre-industrial stage of human populations when birth and death rates are similar, keeping the population relatively stable at a low level. This stage describes the situation our hunter-gatherer ancestors experienced for hundreds of thousands of years. In the next stage, agriculture and the associated urbanisation, the so-called urbanizing/industrializing stage, delivers improvements in health through improved nutrition, medicines, investments in sanitation and infrastructure that bring a drop in death rates. Birth rates remain roughly the same during this stage, so population size increases. When fully industrialised economies emerge, death rates continue to decline in the third or industrial stage, and whilst birth rates start to slow down, the population rises exponentially. In the post-industrial stage, the population growth begins to level off because birth rates have reduced enough to match death rates. In the absence of immigration, some populations even start to shrink as the birth rate drops below the replacement rate of 2.1 children per woman. These stages make up the demographic transition model.

Much of Africa and parts of Asia are still in stage two or three, the exponential growth phases of the demographic transition. The UN estimates that the population of Africa may reach 2.5 billion by 2050, accounting for a quarter of the global total, and continue to grow to nearly 4.5 billion by 2100, which will become about 40% of the world's total. The population of Africa first surpassed one billion in 2009, with a doubling time of 27 years on a growth rate of 2.6% p.a. If the transition model holds for Africa, then these numbers are daunting. The transition needs to be as rapid as possible if any of the resource bases can persist.

The demographic transition model argues for economic development to help reduce crude death rates. Then, improved health and well-being through access to medicines, safe drinking water and sanitation, and information about the disease allow people to change their decision-making and behaviours to have fewer children. Large parts of the world are now in stage four of this model, and countries that wish to increase their population do so through immigration.

Stage one was sustainable but always challenging due to high birth and death rates. Stage four is potentially stable but freaks people out because of a costly and unproductive aging population. They worry about the state of their economy given that the current economic model, which has been peddled for decades, is that the economy has to grow constantly. We have shifted our innate desire for more-making from reproduction to economics. Every day of the week, there is a pattern in the high street—see a resource; use a resource. Shoes, golf clubs, beer, perfume, handbags, brushless drills, phone case in sparkly pink... You get the idea. We are all guilty. The reality is that nobody needs a four-bedroom house for two people. Nobody needs two cars for two people or four cars in a family of two adults and two teenagers. And yet, this is where stage four of the demographic transition takes us to a level of resource use that is considered expected. And yet all the research says that this 'wealth' from resources does not make us happier or improve our well-being. In most studies, wealth acquisition and aspiration for wealth increase stress. Not to say that the 3 billion or so who aspire to greater wealth just so they can have a washing machine are happy, either. There is a balance point, which is that we shoot way past or stay way under. Meanwhile, the global population grows at 8,000 per hour 24/7. Population may be a taboo subject, but the psychology of growth is welded onto the daily lives of the fortunate and those aspiring to be.

Human population has everything to do with rarity. It has everything to do with rare species and everything to do with rare habitats. Population impacts the fundamental resource base, and everything is about nature providing us with resources, whether to overshoot the undernourished or be somewhere in that sweet spot inside Kate Raworth's doughnut. Where our food, health, and well-being are in balance for the global population. Instead, many are consumed by the fear of the extinction of rare things.

According to UN-Habitat, as of 2018, about 1 billion people, or 24% of the world's urban population, lived in slums or informal settlements.

Writing this chapter makes me incredulous. What is it that we cannot see? Well, good luck with that, the pessimist would say. People are selfish and will always run with resource use triggered by needs encased in their limbic system. Basic biology is a hard act to kick, and it makes sustainability impossible.

Ah, yes, the unspoken ideology that has simmered beneath the surface of this narrative. As the definition of sustainability tells us, *'meeting the needs of the present without compromising the ability of future generations to meet their needs'* typically under the three pillars of profit, planet and people. A definition of sustainability that is impossible for humans. As a highly successful species, we use resources, both natural and man-made, all the time. Overshoot day is sometime in August, meaning we must spend nearly half a year overconsuming renewable resources.

Any single action might, on its own, be sustainable. Always take a reusable mug to the coffee shop. But this does not mean that the supply of expertly roasted coffee beans and milk or the power to run the espresso machine is sustainable. The first steps for individuals might be to change light bulbs, consider diet choices, install solar panels, and drink coffee from a reusable mug. Maybe even have a chat about sustainability with fellow caffeine connoisseurs. It will also make a difference if personal investments are made with sustainability in mind, along with big purchases and life choices. Individuals can push up into societal systems to create pressure for change. They can even shift to a plant-based diet. All these will be good; indeed, they are essential. Except that on their own, they are not enough. Something much bigger has to happen.

Awareness is a start. We can keep the ideas around threatened and endangered species and listen to Sir David and his talks about extinction and the severity of the mass extinction crisis. We can continue to set aside areas for conservation and manage nature reserves and wildlife areas more actively so that they can be more like wildlife and less like parks for goats and introduced pests. And understand that there is a balance between how we utilize the land and plan for the conversation outcomes we expect. Amidst this education can be a shift in thinking from objects, Mary-Lou and her Sandhill Crane, to ecological process.

Triage will be inevitable because we cannot save everything. Decide what species and habitats can look after themselves, what to put effort into saving and what we might have to sacrifice because it's too far gone. This will be very challenging, especially when confronted by the forward-facing eyes. Painful decisions are the reality of 8 billion souls, an inflated species extinction rate, and a land surface already converted to support human enterprise. A triage approach also forces learning about species lost and making decisions about the ones we want to keep through actions that are within our control. This effort is really about managing the conservation estate and agenda and accepting that it is impossible to save all the rare things and that saving them is all we need to do.

Awareness and triage are not extreme technical challenges. We already know most of what we need to make evidence-based decisions. We can measure or model landscapes to locate areas with stable natural capital and those losing that capital due to current or historical practices. We can also predict the land management changes that will maintain natural capital for longer. Often, all that is needed is to look long and put an ecological timeframe on the problem. Spatial analysis and environmental information are available at levels of detail that we never thought possible. So, we could be very precise in space, and with our process modelling, we could be very precise in time. We know how.

However, no technical solutions will mean anything unless people are up for a change. A global public realisation of where we are, how we got here and where a fossil-fuelled six-continent food supply chain is taking us is a start, but it can't change our instinct to make more. Awareness and triage will not get very far without understanding our psychology. That will require extraordinary leadership and a wave of personal responsibility, which I tried to explain in *Missing Something*. Scary structural and societal level changes are hard for an individual to grasp.

The idea of small efforts that can build into larger, more meaningful change has been around for a long time. Think global, act local is a pretty good adage. The phrase's first use in an environmental context might have been by David Brower, founder of Friends of the Earth, as a slogan for the organisation when it was founded in 1971. It could also have been coined in 1977 by René Dubos, French-American microbiologist, experimental pathologist, environmentalist, humanist, and winner of the Pulitzer Prize for his book '*So Human An Animal*'. It might have been Canadian "futurist" Frank Feather who chaired a conference called "Thinking Globally, Acting Locally" in 1979 and has claimed the expression. Either way, the idea has been around the block.

At its inception, 'Think Global, Act Local' was about empowerment, a response to the endless stream of bad environmental news that bewildered and overwhelmed people. Concerned individuals could begin to see how their own efforts could make a difference against the global economic juggernaut because there is more than the sum of the parts. However small, each part matters. 'Every little helps' and 'Look after the pennies and the pounds will take care of themselves' are hewn from the same psychological rock.

I like that. You probably do, too. Whilst at some level, we all know that we are small and insignificant, it is a fact that our egos refuse to accept. So anything that implies greatness, even via aggregation with our fellows, feels good. Unfortunately, most individual acts are actually insignificant against the tide of economic development. Standing down the bulldozer and chainsaw by living on a platform in a tree at the edge of the forest is meaningful at the time for the tree dweller, the dozer driver and the handful of people following the social feed. But not to the logging company, as their licence will rarely be revoked. In time, the act of bravery and defiance is forgotten, and a new agreement is reached to create jobs for the timber industry and paper for the printer, perhaps from some forest down the road. And yes, the cardboard used for the placards at the demo originated in a tree. The road, vehicles, and fuel that transport the protagonists and antagonists to the forest have an environmental footprint. The tree dwellers' family has jobs in the city, and after tossing their disposable coffee cup into the bin, they went online and transferred $500 into their daughter's account, claiming the gift against tax.

The local act was noble and courageous. It will have raised awareness a little and stalled a poor development decision—for the record, I believe that any further logging of primary forest is not development but degradation of the worst kind and that timber production should be all about revegetation and management of already logged forests—only the act did not go global. The thought might have, but the action did not. Forestry today is as voracious as ever.

Modern food supply chains are long. The average food item in the United States travels about 1,500 miles (2,414 km) from farm to plate—imagine the impossibility of this to a hunter-gatherer living off the habitats he could walk into. Source: Weber, C. L., & Matthews, H. S. (2008). Food-Miles and the Relative Climate Impacts of Food Choices in the United States. Environmental Science & Technology, 42(10), 3508–3513

So here is a suggestion. **Act global, think local**.

At first, this sounds stupid. If individuals feel so small and insignificant, they cannot possibly act globally. It is beyond them and why the original mantra became popular. The best they can do is vote for global change and look where that got us. But they can think locally. What happens if we think about everything we do? Think about the disposable coffee cup, the commute by car for an hour yourself, the printer when there is the cloud, and any number of commonplace actions that all have an environmental cost.

If we think, we question. The answer might be that coffee is an essential that should not be passed over, and, anyway, it was fair trade coffee that spread the love across the world—the thought at least triggered a logic flow. Do this many times, and the logic starts to accumulate. We start to think critically. Gather enough logical thoughts, and the futility of so many of our individual and collective decisions will become obvious. If you need more on how to do this, I have written a guide called *Being a Mindful Sceptic* that uncovers the power of mindful scepticism to answer questions objectively.

Undoubtedly, environmental issues are the aggregation of all our common actions. A world with over a billion wealthy people and another three billion more hot on their heels will strain the limits of natural resources and global resilience. And changing the light bulbs is never going to be enough.

If we think locally, we become more aware. We start to realise the extent of the challenge, and only then does acting make sense.

A statistic bandied around is that in mature economies, the food component of the family budget is usually around 10%, slightly less in the US and slightly more in France. Not all families are at the average—medians would be a better metric—with the poor spending much more of their income on food, but if food prices doubled, most people would find a way. We'd complain, and it would be unpleasant, and there would be lots of people ignoring the politician's claims and having good arguments about the increased price, but sooner or later, we'd find a way, and we'd be able to spend 20% or more of our budget on food. We know this because most of the rest of the world in the emerging economies spend much more of their available income on food. The average Ugandan, Indonesian or Peruvian family will spend 30 to 50% of their available income on food. That means if food prices double, much of his existence is threatened. It explains why the cost of food is one of the reasons subsistence agriculture is still prevalent worldwide. People can't afford food, so they must grow it for themselves. Every meal they don't have to buy is money they can spend on housing, health, and education essentials.

Over 500 million subsistence farmers operate around the world. Raising the world's wealth is assumed to reduce reliance on the subsistence system and improve its production efficiency whilst retaining nutrients so that we don't have to cut down so many forests for the subsistence rotation to continue. Pressure on the Amazon includes the needs of smallholders for land to grow food for their families. It may be facilitated by the large corporations putting the roads in and flattening the landscape to put in their palm oil plantations and their beef ranches. However, around the perimeter, that access allows smallholders to take advantage. And if you were in their position, you'd do the same.

What questions must I ask myself to come to terms with these realities? How much of the change is about me as a person? As much as it is about structural change to the way that we grow and consume our food. Ultimately, that's where it all lies: how important we think this is as individuals. Do we believe there needs to be a change in how we conduct business? Or is it perfectly fine to be self-orientated and self-centred about these matters? Thinking locally, act globally?

Sustainability is a collective outcome. It's not really about individuals, even though individuals will contribute. And we must be honest with ourselves. Changing the light bulbs and turning off the air conditioning is all good, but we can't say that we've saved the planet, particularly if we haven't addressed the item on the topic of this book. We can't give $10 to the Koala Foundation and believe we've done our bit because that is not even the problem. Awareness is important; no more falsehoods. We must come to terms with the truth of what has happened—the fossil fuel pulse has been converted to 8 billion people eager to make more. Awareness requires knowledge, understanding, and a certain skill set that can be learned. Most of it is available for free on your smartphone.

More difficult is the honesty required to accept the truth. Eight billion people require enormous food, clean water and a place to live. Nutritionists say the average male human needs 2,500 kilocalories daily for maintenance, and the average female slightly less. The average American consumes nearly 4,000 kilocalories, and people in Australia and the UK consume 3,500 kilocalories. Each person in those countries consumes a thousand calories a day more than they need to be healthy. Calories are not quite a good measure of consumption because nutrients are as important as the energy component of food. So we could do the same sums for nutrient content, but we'd find that most people in the developed world eat more food than their bodies can absorb. But even this is not the real challenge.

I was born in 1961, a vintage year that gave the world Julia Louis-Dreyfus, George Clooney, Michael J. Fox, Eddie Murphy, Woody Harrelson, Meg Ryan and Diana, Princess of Wales. The global average energy from food per person was 2,181 kcal, and there was a tick over 3 billion people alive, generating a daily demand of 6.7 trillion kcal. Jump ahead 60 years to 2021, and the average daily food consumption had risen by 778 kcal, the equivalent of a large cheeseburger with fries from a fast-food restaurant, a plate of spaghetti carbonara, or three chocolate chip cookies and a large glass of whole milk. Population had grown by 4.84 billion for a 250% increase in the total daily calorific demand to 23.4 trillion kcal.

Global wheat production in 1961 was 222 million tonnes. One tonne of wheat generates roughly 3 million kcal of food energy, so the world needed roughly 2 million tonnes of wheat per day to meet food energy demand, 730 million tonnes for the year. Wheat could supply roughly 30% of the demand. Luckily, there was more than bread on the table. By 2021, global wheat production had increased to 773 mt, roughly a 2% annual increase over 60 years. Meeting the global demand for food energy from wheat in 2021 required 7 mt per day, 2,555 mt per year. Wheat could still supply roughly 30% of the demand. A simple check on this calculation is easy enough—there are more people daily.

Meeting the needs and wants of 8 billion people and counting is not why there are rare things; nature generates rare things as easily as common. People are the reason for faster loss rates, unsustainable resource use, and a planet-level disturbance sufficient to trigger a mass extinction. These people have rights, and access to food is one of them.

The United Nations Food and Agriculture Organization (FAO) estimates that global wheat production typically varies by about 3-7% from year to year, with a general upward trend from improvements in farming techniques and technology. In recent years (2010-2022), global wheat production has ranged from about 650 million tonnes to 780 million tonnes per year

Universal Declaration of Human Rights

Article 3 | Everyone has the right to life, liberty and security of person.

Article 25 | Everyone has the right to a standard of living adequate for the health and well-being of himself and of his family, including food, clothing, housing and medical care and necessary social services, and the right to security in the event of unemployment, sickness, disability, widowhood, old age or other lack of livelihood in circumstances beyond his control.

Chapter Nine

Saving a loaf of bread

In the shopping malls and high streets of Australia, I will likely be accosted by a passionate advocate wearing a bright-coloured t-shirt, a lanyard for authenticity, and a wide grin. She carries a white bucket only partially obscured by a sticker for whatever foundation or conservation organisation she represents for the day. The faux polite and usually young individual who is sometimes a volunteer and sometimes paid for her time will ask me for a donation to save the koala, just a few dollars.

"But I don't carry any dollars," I say, " and I haven't for years."

"No problem. Here, we have a mobile point of sale; you can tap for a few dollars. It will save a koala."

In my mind, I have two questions: How and why?

How can a few dollars be spent to save a koala? What exactly is your organisation going to do with my money? Presumably, some of it pays for your time, your t-shirt, the many ads your folk run in the media, and the overheads to keep such a campaign alive. The rest goes to what? Planting trees, rescuing injured animals, putting up water points in the bush after wildfires... perhaps the preparation of erudite submissions to government committees. Knowing what happens to my donation when it leaves my account would be good.

Then, why should I pay? What evidence can you give me that your slogan 'Save the Koala' is necessary? The koala is one of the hundreds of thousands of animals at risk of extinction; even humans fall into that category, for goodness sake, but somehow, the koala is unique. And anyway, aren't there koalas everywhere? I think I read there are populations from the Atherton Tableland west of Cairns in Queensland to islands off the coast of Victoria and South Australia in the south, and west to central and western Queensland, New South Wales and Victoria. The buggers are everywhere. Are you expecting all their habitat to be destroyed across a land area bigger than most countries? I need you to tell me why they need saving and from what?

In the real world, I said, "Thank you, but not today."

The look the youngster gives me from behind her smile is one of disdain that says you are one rich white f*cker, so privileged you can't even offer a few dollars to save a koala; what a complete bastard you are.

I smile back and try not to let her see my anger.

She turns to the next shopper, who eagerly taps their phone on the mobile point of sale.

The truth is that '*Saving the Koala*' is a ruse, an obfuscation, a means to deflect attention from the actual problem. It is clever because it is emotionally charged, and that phone tap will send an endorphin shot to the brain of even the harshest sceptic. I would even feel good about it. Saving the koala or the panda or polar bear or any other creatures with fur used in these extinction statements is all because, to humans, they look cute, a bit like cuddly babies. We like them a lot. And when we want something or someone, it turns off many mental filters that protect us from scams and charlatans. The cuteness suckers us into many an irrational response.

Australian Charities and Not-for-profits Commission (ACNC) estimates that Australians typically donate around $12.5 billion to charities annually and roughly 2.4% of this goes to the environment and conservation sector, roughly $300 million.

This ruse is enough to turn me off all the green movements, for they all do it to some degree. They play on our emotional responses to further their agenda. But before I go on or push you away, I am not saying that the extinction of the koala or the polar bear would be desirable. The loss of biodiversity is a huge global challenge that must be addressed with all the will humanity can muster. And even though it's just one species, the koala represents biodiversity, the diversity from billions of years of evolution that humans are dismantling in a few generations—it matters little if they are rare or common.

But there is another question about what is being saved, the specific animal or the species. Do the funds go to specific critters or actions for *Phaccolarctos cinereus*, the Latin binomial that names the species under which each koala is classified? Most who donate have never considered the distinction. They assume that the money will mean **a koala**, not **the koala**, one with fluffy fur scorched in a wildfire, will be looked after.

Perhaps it will be bandaged and cared for until it is well enough to return to the wild. Of course, conservation organisations know the difference.

Here is the plea from WWF...

> *Now uplisted to ENDANGERED status along our east coast, they're at a critical crossroad and urgently need your help. Without stronger nature laws and more places to call home, koala populations are on a slippery slope towards extinction by 2050 in eastern Australia. There's no time to lose.*

A similar one from The Koala Foundation encourages us to save a koala but states its core objectives beyond the individual koala...

> *Our main goals are to refine and update the Koala Habitat Atlas mapping of all wild Koala habitat and to get the Koala Protection Act enacted to ensure Koalas and their habitat are protected.*

It turns out that the koala is unlikely to go extinct anytime soon, perhaps not even in eastern Australia. Evidence for its extensive geographic range, multiple populations, pockets of high abundance, and ecological niche that covers disturbed and regenerating eucalyptus woodland suggests that the species is neither rare nor at any greater risk of extinction than any other terrestrial mammal that is not a human or livestock. I go into the details in my book with Chris Scott, *The Mindful Sceptic Guide to Saving the Koala*. But here, before a heated argument builds over the relevant facts, risks, and values, there is a need to pause.

What exactly is the koala, *Phaccolarctos cinereus*? Throughout this book, I have used the word species as though it was obvious what it meant. The rare things are species. But what exactly are they? The textbooks will tell you that a species is a fundamental unit of biological classification that refers to a group of living organisms that can interbreed and produce fertile offspring. Species are considered the most basic unit in the taxonomic classification system, the unit below the genus level, and each species is assigned a unique two-part scientific name, for example, *Homo sapiens* for humans and *Phaccolarctos cinereus* for the koala.

Organisms within the same species share many similarities, including genetic makeup, physical characteristics, behaviour patterns, and ecological roles. Members of the same species can interbreed, mate, and produce fertile offspring. This is known as the *Biological Species Concept*. Individuals within a species share a gene pool and can exchange genetic material through reproduction. They generally share similar physical traits, such as body structure, colouration, and anatomical features. They occupy a specific ecological niche, sharing similar habitat requirements, food sources, and interactions with their environment.

As Darwin and Wallace determined, a species represents a unique evolutionary lineage, descending from a common ancestral population and sharing a unique combination of genetic characteristics. A species means something as a unit of organisation and for understanding biodiversity, studying evolutionary processes, and managing conservation efforts.

It might be surprising then to learn that the definition of a species is a subject of ongoing debate and discussion among biologists, as different concepts emphasize different criteria.

The *Biological Species Concept* is the most widely accepted and traditional definition that works best for the bigger, sexually reproducing animals and plants. It defines a species as a group of interbreeding or potentially interbreeding natural populations reproductively isolated from other such groups. This concept emphasizes the ability to mate and produce fertile offspring as the primary criterion for delimiting species.

Then there is the *Morphological Species Concept* that defines species based on their physical or morphological characteristics. Organisms with distinct structural or anatomical features are considered part of the same species.

Suppose you prefer to follow the genetics and the evolutionary pathway. In that case, the *Phylogenetic Species Concept* defines a species as the smallest group of organisms that share a common ancestor and can be distinguished from other such lineage-based groups by a unique combination of derived traits. Or you might prefer the *Genetic Species Concept*, which defines species based on the presence of distinct gene pools or genetic clusters. Organisms that share a significant portion of their genome and have little or no gene flow with other groups are considered separate species.

If you are more ecologically inclined, the *Ecological Species Concept* emphasizes organisms' ecological niches and adaptive traits. A species is a lineage or group of organisms occupying a specific ecological niche or adaptive zone.

It is enough to give Eric the willies. Was his White-rumped Swift, *Apus caffer,* a vagrant, a misidentification of the common swift *Apus apus*, or not even a species at all? While the species concepts overlap and can be complementary, they may also lead to different species delineations in certain cases. The choice of species concept may depend on the specific research question, the organisms being studied, and the available data. In practice, a combination of multiple lines of evidence, including genetic, morphological, ecological, and reproductive data, is often used to delineate species boundaries. All Eric had was the view in his very expensive binoculars.

Unfortunately, there is confusion about species. This entire narrative about rare things assumed that you knew what I was talking about—species. We need a unit of organisation for all the various uses. We protect species for their inherent value, assuming each species is a unique product of millions of years of evolution and has an irreplaceable role in the natural world. The presence, abundance, and diversity of species can indicate an ecosystem's overall health and integrity. Monitoring species can provide valuable insights into the impacts of environmental changes or human activities. We assume that each species plays a specific role in its ecosystem, such as pollination, seed dispersal, nutrient cycling, or regulating population sizes of other species. Losing a species can disrupt these ecological processes and have cascading effects on the entire ecosystem. Many species have actual or potential value for humans, including food sources, medicinal resources, or sources of valuable compounds or materials. Species are the repositories of unique genetic diversity, representing millions of years of evolution and adaptation. This genetic diversity is crucial for species' long-term survival and adaptability in the face of environmental changes. Species hold cultural, spiritual, or aesthetic significance for various human communities, contributing to our diverse cultural heritage and appreciation of nature.

They had better be real, or a heap of effort would have gone down the gurgler, not least Eric's list of sightings. But we must remember that what we are saving, what we define as rare, is an abstraction, a made-up entity that conforms to some rules invented to help interpret evolution. Species might not be. Yet they are what is lost—a koala dies, *Phaccolarctos cinereus* goes extinct.

Nobody knows how many Koalas there are in Australia, not even how many populations. The facts are a known unknown.

Before I can build a more positive ending for a torrid journey about the rare things, I will tell you about an idea that came to an ordinary gentleman one sunny morning in response to the same urge that has good folks ready to tap the point of sale in the shopping mall.

If I may, I'd like to tell you a story about George, a forthright and well-meaning man, sound of heart and mind, who had a problem. The problem was this. George desperately wanted to conserve a loaf of bread. It is hard to understand why he wanted this, but he did. He wanted to save a loaf with a passion that made his stomach ache. Now we are talking about preserving, keeping, protecting, looking after, and perhaps even hugging this everyday product. Clearly, bread is for eating, not preserving, but George just couldn't help it. Something deep inside told him he had to preserve this loaf.

This thought began suddenly one fine spring morning. Having bought a delicious fresh loaf of bread from a fine baker on the village high street, George had this feeling that although he himself was in tip-top shape, this might just be his last loaf. It was quite a surprise but it gripped him like an attack of cramp. Maybe, just maybe, this loaf was the last of its kind.

George was overcome with a profound sense of duty to protect this oblong cob, this last loaf. Like some great turning of the tide, it was inevitable. He had to do it. And yet, he found himself in a quandary. The desire to protect was powerful, dutiful, and noble, yet nutrition triggered by a little hunger was why he purchased the bread. Part of George wanted to eat a slice or two of the loaf with peanut butter and jam. This was, after all, a very fine loaf baked to be eaten with peanut butter. And here it was, in his hand, at its very best, soft enough to squeeze between the fingers and with firmness that only the crispest of crusts could support. It even had the warm smell that first danced across his nostrils when he entered the bakery and was about to infuse his clothes and the deepest recesses of his Nissan pulsar.

Perhaps if I was frugal, George thought, *I could eat just one slice a day, and the loaf would last a couple of weeks*. This seemed like a reasonable plan for Monday and Tuesday, a bit hard on Wednesday, probably pretty stale by Thursday and altogether too risky to his health by next week. No problem. What, after all, were freezers made for? A freezer was the perfect repository to preserve a loaf of bread. So, our smart man decided that he would pre-slice his loaf and put it into his freezer so that his tactic of frugality could be implemented without risk to his digestive system. Sure enough, the loaf would last for a couple of weeks.

Defrosting a pre-sliced frozen loaf is not the same as cutting a doorstop fresh from the cob. You miss the pleasure of rasping the blade across a crisp crust that still holds the amazing soft centre with affection, all set off with an aroma that makes the mouth water and the lips say, "More, please". To pre-slice and freeze was a compromise, yet this loss of quality was acceptable to George, for all he had to do was retrieve a slice from the freezer, thaw for a few minutes at room temperature, apply the spreads and eat.

This was a logical solution. Yet it left George with a niggle that things were not satisfactory. After a week or so, the loaf is finished. No more loaf. If this was the last loaf ever, surely there was something better he could do than eat it. There should be some record of it as the last ever. Something to remember it by, a gravestone perhaps, or something more direct like the depressions in the cushion of grandma's favourite armchair. Perhaps the paper bag wrapper that kept the loaf clean in transit and moist for its natural life could be framed and hung in a local museum.

George fumbled for options until one came to him in a surprising moment of clarity. *What if I preserved the loaf indefinitely?* A big grin lit up his face. What an idea. "I am a walking genius," he said to nobody within earshot.

The frozen option would do pretty well so long as the freezer worked and the power stayed on. The loaf would keep for a long time, perhaps forever. Sure, another food source would have to be found, and luckily, there were some handy rice crackers in the cupboard. You know, the ones that line the calorie-counting shelves of the supermarket and taste like cardboard—at least what George imagined cardboard would taste like. But this was a small price to pay for preservation permanence.

George made a decision. He must get serious about this cryogenic method, and if he did, if he took a few precautions, he could indeed keep his loaf for what he thought would be forever. So he bought a second freezer as a backup in case the first one broke down and installed an alarm system to tell him if this happened. At the hardware store, he bought an uninterrupted power supply system and, in case the power went off, their most expensive portable generator. Two jerry cans of petrol for the generator were stored safely in his garage. He toyed for days with the idea of buying a gas freezer, just in case, until he had to get one to stop the worry.

Fearful of risk and natural disaster, he moved his new preservation setup into a facility he had built in his yard. Made out of brick and clad with the latest fire retardant, it was a very solid structure that included the latest surveillance cameras, a digital alarm, an anti-tamper system and all the power backups for the electronics. Standing at the top of his yard and admiring his museum, George figured he had the bases covered. A mate of his, a risk assessor for an insurance company, agreed. So, barring acts of god or terrorism, his loaf was secure. George had made the loaf safe, and our man was pleased with his work. He could not eat the loaf, having gone to so much trouble to keep it. But the frozen loaf was preserved.

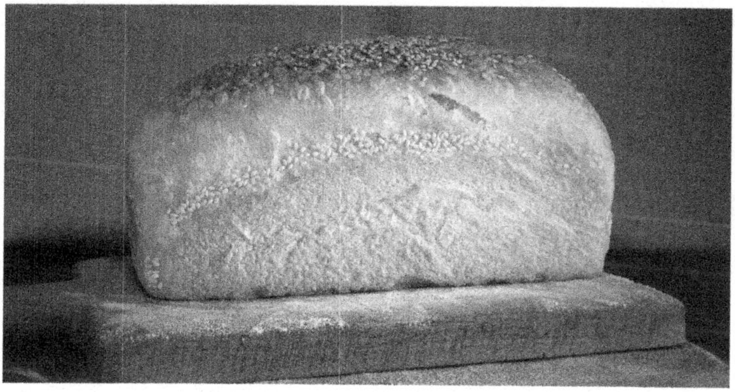

Many months passed, and George began to forget about his loaf. He only checked on the facility occasionally because he was sure that the electronics were watching. Sometimes, he thought about the loaf in an idle moment at the office and was glad he had installed such reliable surveillance, but by the time he got home, the loaf was forgotten. He did show the loaf to his young nephew when he came to visit, although the little kid, who was still a year off kindergarten, didn't seem impressed and kept asking why he couldn't eat a slice of the loaf with peanut butter. George put this down to the innocence of youth.

But as the years rolled by, George began to think that his loaf was important. It was now 15 years old, which is quite old for a loaf of bread, and on the outside, at least, it looked pretty much as it did when first frozen. George's choice of a modern frost-free freezer was a smart play, as was the idea of not opening the fridge unless absolutely necessary. He began to imagine how valuable this bread would be to his nephew when the youngster was old enough to appreciate an old loaf. Indeed, it would surely be worth a lot of money. He had read in the paper that a young lad was given a bag of coins by an aging aunt that included a very rare dragoon worth thousands. George imagined that he could leave the loaf to his nephew in his will, but if he did, then there would be things to do about the long-term storage of his loaf. How should he ensure its preservation until his nephew was old enough to appreciate it? It might not be easy to find someone with enough passion to step in and take responsibility for a frozen loaf of bread should anything happen, like being on the wrong side of a moving bus. He had gotten the loaf this far and wanted it in the family, but it had to be safe. He didn't want to risk the impetuosity of youth.

He called his lawyer.

"In the future," George began, "when wheat blight has made flour scarce, as it surely will, bread will become rare. My loaf will be a national treasure. It would be the only intact, albeit frozen, loaf of bread left in the world. The only remaining example of a food that half the humans in the world used to consume every day." George paused. "There will be some value there, surely. Maybe even a way to cash in." It was in the nature of the profession not to question a client's motives, and the lawyer calmly offered the solution. He would gladly establish a trust responsible for the preservation of the loaf. George accepted eagerly. The firm would prepare the papers.

Although he had the legalities covered, thoughts of future rarity played a lament in George's head. He still felt uneasy about his frozen loaf; there was something about keeping it in its cryogenic state indefinitely. *Why would that be good?* he thought. There is no nutritional benefit; folks only get to look at it occasionally, perhaps only on television or the newspaper's inside pages.

Like things that happen after a couple of glasses of wine on a balmy summer evening, George asked himself some important questions. *How did my loaf of bread come into existence? Why is it as it is? What am I doing keeping it past its sell-by date?* And many other tantalizing teasers that were probably just the wine talking.

Still feeling hungover and uneasy in the morning, George sat at his computer. He enquired to the ether about the lost art of bread making—a lengthy search uncovered details from some old folk who remembered what to do. Flour, water, yeast and salt are mixed in the right proportions to make a dough, then some mechanical energy is applied with healthy kneading, and the dough is set aside to rest while the yeast does its thing. Bake the risen blob in an oven for a while, and almost by some miracle, a crust appears around a soft, aromatic centre.

George also found that you could grind a wide range of grains into flour to produce bread with different textures and smells. It was even possible to add other ingredients for flavour, colour, and nutrition, and voila, you had a whole range of breads. You could also mould the dough into a limitless array of different shapes before the miraculous intervention of heat to create baps, loaves, sticks, cobs, rolls, bagels... anything.

The most important thing that George discovered was that if you have the ingredients and an oven, plus some ability to follow the basic method, you can always make bread. George imagined that it would not take much to return to his initial plan: to have enough bread to keep and eat. If more bread was being baked, George realized he could eat a few slices daily with peanut butter and still conserve his loaf. Not the loaf, but at least a loaf. Every few days, there would be a fresh one to slice. It was very satisfying, not least because the loaf would be doing what it was made to do and not just stay in the freezer using up space and energy.

Just as George was about to abandon cryogenics, he felt a twinge, that nagging feeling in the pit of the stomach, something remarkably like worry. *What happens if the ingredients run out?* There might be a day without bread. George reassured himself that this would be fine so long as the baker could bake tomorrow. A loaf would be edible for at least a few days. But what if it takes many days to get the ingredients? There could be no loaf baked for a week or longer. This situation felt risky and violated the constitution of the trust. Conservation of the loaf with the baking plan required that the process of bread making was secure. The baker needed an uninterrupted supply of flour, water, yeast, and power. The oven needed to fire every day. George knew that he must learn more about the ingredients needed to make bread, including where they come from, what happened to get them into the right shape for baking and how they got to the baker.

Several days later, George had only figured out the local flour supply line. It was long and involved many more people than the farmer who, thanks to a rural heritage grant, had begun to grow wheat engineered to resist the blight. A dozen or more folk were involved in turning the kernels into flour and getting the right mixture into the oven, and with more people, skills and crafts involved, it was obviously more complex to secure supply. It was one thing to know the supply chain and quite another to be in sufficient control to guarantee an indefinite supply of all ingredients and skills to use them.

George had several restless nights. He was now quite concerned about the long-term future of his loaf. His plan to return to baking looked shaky when he was awake and became a nightmare in the dark. A small fortune was needed to ensure supply from the power corporation, the wheat farmer, miller, microbiologist and salt distiller, not to mention a cut for the many companies transporting these goods to the baker. A hefty lottery win would not be enough to buy all the needed land, equipment and expertise. The process was too complicated. It was not complex, but complicated. The freezer option might be the engineering solution, but it was much simpler.

George agonised for the rest of his life. More years passed, and he gave up on the resources to secure his bread-making supply chain and resorted to the frozen option to preserve his loaf. In this, at least, he succeeded. The loaf had been frozen for 34 years when George died. It survived another decade under the trustee's care until a share market crash sent the trust fund bankrupt, unable even to pay for the electricity to run the fridges. When the Trust was wound up, no one believed that the loaf of bread had been frozen for longer than anyone could remember.

Wheat blight had eradicated flour from the human diet several decades earlier. The bakers sons were forced to other grains and finally figured out how to make something edible from maize meal. Bread made from wheat became ancient history. Only the old folk remembered what bread tasted like. One ancient gentleman, apparently old enough, was asked if he recognised George's frozen loaf as bread, but he said it wasn't twisty enough. The bread was always twisted in his grandfather's farmhouse. So, it was decided that George's loaf was a fake.

There were portable DNA fingerprinting machines, but somehow, nobody bothered to check for wheat genes in the loaf. It was as though the passing of time had completely devalued this, the last loaf. No one remembered what an old-fashioned loaf of bread was like, and, worse, this memory loss didn't seem to bother anyone. Like the dodo, no one had ever seen or experienced its value. Only the historians had any interest left.

George's nephew had become a civil engineer and enjoyed fortune and considerable fame as the founder of a company that revolutionized the design of desalination plants. He occasionally recalled the ingredients for bread, something that his uncle used to tell him when he was a kid, but he couldn't remember why.

Chapter Ten

Hiding behind the rare things

Most of this book has been about why the rare things are not the only thing we should concern ourselves with when it comes to conservation, protection of the environment, unsustainability, and the likelihood of human beings persisting through what we hope is an orderly demographic transition. The truth is that numerical and technological success on the back of fossils has got away from us.

In escaping the Malthusian trap, we have made a gigantic mess. It's embarrassing and, for anyone paying attention, terrifying. A common response is to hide behind rare species. Limbic responses to the fear trigger instincts that make us feel better when saving the Koala, the Cheetah, the Ghost Orchid, or even a loaf of bread. Feel good is salable and makes it easy to get people excited about saving iconic species... and neatly avoids awareness, which I suggest is a crucial obligation of any *Homo sapien*, wise man.

Instead, the rare species trope has stuck. In part because it benefits advocates on both sides of politics. Species conservation can be framed to build support across the political spectrum. For the political left, saving the koala can mean environmental protection, preserving species for ecosystem health benefits of biodiversity, sustainable development, social justice and public health. For the political right, conservation can be framed as job creation in ecotourism and resource management, the protection of national identity and cultural heritage, respect for property rights, the long-term viability of hunting and fishing, and even national security through food security and provision of natural resources important for various industries. In the centre ground, conservation efforts can be spun into intergenerational equity, cost-effective governance, technological and scientific advancements, international cooperation and local empowerment. All potent stuff, fuelled by the trigger of those doughy, forward-pointing eyes.

True conservation should treat each species equally and maintain the diversity of life as a whole. Many argue that saving iconic species is an umbrella for all the other species. But clearly, that's not happened. Biodiversity loss continues at a rate well above the background rate, and the latest response to some of the solutions is that we'd need to put two-thirds of the planet under reserves to save what's left. This is impossible without destructive food shortages.

Human numbers have reached a point where the choice is no longer both but one or the other. Either we have the human population as it currently stands and continues to grow, surviving and not killing each other—in other words, providing enough food for a certain amount of stability across the planet—or we save iconic species at the expense of that food production.

Now we know which way that argument is going to go. Fundamentally, the individual comes first. Very few people are able to sacrifice themselves for the greater good. We give them awards and hold them up as saints when we find them. The majority of us don't have the courage. We're much more concerned with not making choices about anything in particular and living our lives vicariously through our phones.

So, here is a subversive thought. The conservation movement is to blame for failing to save the rare and endangered. They've latched onto the rare species to sell what they believe is a difficult message to be sold in any other way. They appeal to our base instincts rather than to our higher cognitive selves. We have educated our youngsters in a few things in this book besides the rare species. If you asked the average high school student what they know about soils, their replies would be short. On television recently, an Australian high school kid sitting with her Gogglebox family thought that Mexico was in Europe. I don't know what we teach them in schools, but it's not what they need to know to make sensible choices about their grandchildren's future.

The right and left of politics are just as responsible for this failure. Because they have lapped up the fact that the greens have chosen the rare species route, they know that they can play lip service to that and put money into iconic circumstances and certain places, the barrier reef, for example, or the Galapagos Islands or rhinos in southern Africa or icons that people can get their head around and see that something has been done. Meanwhile, policies on land management, land protection and soil management slip away to be decisions made by individual landholders who are plagued by survival and profit more than they are in the long-term performance of their landscape.

The abject failure of conservation is a brutal admission. It is one that many people would rail against, and I expect to be trolled. Only it's a statement we have to hear. That all sides of politics are not in our best interests when it comes to the environment. The left wants to conserve the wrong things, the right gives lip service to that whilst at the same time slackening off policies to allow individuals to have dominion, and the greens are in la la land.

And, of course, all sides play the religious card to justify their positions. The church also has some deep soul-searching to do on this matter. I grew up in the Salvation Army and was steeped in its bizarre doctrines, constructed towards the tail end of the Victorian era and designed to save souls from the devil. But nowhere in those doctrines does it say anything about believing in the environment and nature and nurturing the natural world that brings peace and understanding. I think I became an environmental researcher partly because my subconscious said well, this is not right, surely. An all-powerful deity would be keen to keep what was created rather than let one organism trash it. Of course, I don't believe this, but if that's the doctrine presented, you would imagine that the all-seeing almighty would be keen to have his creation looked after. Unless he did design it for us to have dominion, in which case you might as well put the book down and get back to it.

In decision-making situations in our society, culture and religion, the environment is, at best, an afterthought, or it's been bastardised to promote certain opinions. But it all comes back to us as individuals. We allow and facilitate the numpties.

My former mother-in-law was very keen on cats. She decided that she would contribute to the world by saving them. We're talking about domestic cats, including any cats she found on the street, many of whom were already feral. She enticed the animals with food and then brought them back to her house. They lived inside in their dozens, imprisoned and never again able to go outside. Needless to say, her home was not a pleasant place to visit. Each cat had marked out a tiny square foot of territory and spent most of its day protecting that territory from all the others.

A house full of feral cats caricatures the warped sense of what we're supposed to do in all this. We do not know what we are, what our place in nature is, or what our story should be. There's an excellent documentary that came out recently of a South African man who was struggling with mental illness. He was fortunate enough to live by the ocean and chose to combat his depression by snorkelling every day. Remarkably, he built a relationship of sorts with an octopus. It's a touching story about how a human being can associate with nature in such a way that nature associates back. It recognized him as part of the environment; in his own words, he felt part of the place rather than a visitor. This is remarkable, considering it was a kelp forest he visited daily.

We must reassess how we connect with nature and what nature means to us. No more of this left and right politics nonsense using threatened species and rare species as an excuse to carry on doing whatever it is you want to do. I think that it's time to access more than the rare species. To remember that nature is the supplier of our well-being. It's the supplier of all our basic needs. And that won't change even with our extensive and increasingly complex technologies. Nature will underpin our existence for as long as we are on the planet. And if we carry on the way we're doing, not recognizing that fact, ignoring the really important matters—the more than the rare things—then the ability of nature to support us will continue to be eroded.

And as I'll remind you, nature doesn't care about that. Nature will kick on regardless and pay no real attention to us. So, our importance is not that we are dominating nature or destroying it. It will bounce back, given enough time. That problem is ourselves. We are the ones who have to recommend and obtain another relationship with the world and with each other, which means we can ask these questions intelligently and think through answers. And put into policy and practice a whole new way of looking at the world.

We can turn around the 40% of the soil that is degraded. We can rehabilitate soils just as we can rehabilitate vegetation. We can even do all that while producing enough food to avoid global instability. If we stabilise numbers, heal soils, and reorder the food supply chain, then we can even have a go at rewilding. But we can't do any of this if our only focus is rare species. If all we do is save the koala by pretending it needs saving and let people assume that all will be well once they've paid their ten bucks to the koala foundation. That level of naivety has to disappear. As Sir David has done in his most recent documentaries, we have to begin to say it as it is.

Humanity faces an existential crisis. We are about to annihilate ourselves as we fall off a resources cliff, even as digital money grows and we consume 97 million barrels of oil a day. And yes, I know, it's been said before. Many people have pointed to the cliff, there it is, and we're going to fall off it, we must change our ways. And they were all right. If we changed how we dealt with things back in Malthus' time or even when Paul Ehrlich wrote the Population Bomb in the 1970s, then some of the problems we are facing would not be or be much easier to overcome. We would have paid attention to soils; we would have paid attention to habitat change. We certainly wouldn't have a climate crisis because that would have been fixed long ago regarding emissions. We would also have developed adaptation technologies to the present climate changes. But mostly, we didn't pay attention, remained blissfully unaware, and hid behind the rare things.

Now, we're at the pivotal point where we can't make any more fuss. We can't make any more false claims. We can't focus on the wrong things. We must focus on more than the rare things.

And here is how we do it.

The next 30 years are critical. The risk to survival of *Homo sapiens* as a species is fixable right at this very moment in time. We have the knowledge and the understanding of what to do, where to do it, and how to do it. In one set of actions, we can shore up food security, reverse our challenges of soil degradation, reduce the climate problem, and recreate a landscape matrix in which we can slow or even stop the current rates of biodiversity loss. And we can do this with just a few relatively small changes to how we perceive and live our lives.

The solutions are simply a choice. We can take them on and do them and, where necessary, force them through our societal and jurisdictional regulation. We can make it part of our culture. And we can do it even by allowing a certain amount of face-saving for those people who have to change and are nervous about it.

All this begins with agriculture—regenerative agriculture, to be precise, growing food to restore carbon and plants to the soil to return those soils to something like their original productivity.

The reason is simple. Plants and soil are integrated through ecology to maintain life on the planet. Modern intensive agriculture replaces reliance on biology, ignores the ecological processes to a greater extent, and makes them input-output. These changes, these simplifications, mean that the systems are no longer viable over the long term. They're not sustainable systems; they are degrading even as we continue to get yield from them. Regenerative agriculture is about returning to the biology of the soil. It's bringing the soil back to life. All it requires is to put plants back into the system to retain cover on the soil, have minimum tillage practices and make the soil into what it was: an organic soup of activity and biological exchange.

Regenerative agriculture is about multi-species, it's about multi-crops, it's about multi-livestock, it's integrating the livestock with the plants, it's moving the nutrients around the system in the way that the original systems used to operate. There is a massive reduction in inputs, fewer nutrients, fewer pesticides, herbicides, and fungicides, and less need for genetically engineered crops because you're not trying to homogenize to maximize yield across every hectare. Instead, the objective is a system of productivity without the need for excessive inputs of energy and nutrients. Specific actions can be nuanced and integrated with intensive production to create regenerative agricultural systems in almost any place on the planet, even heavily degraded areas.

It is estimated that 10% of US farmers are already involved in regenerative agriculture of one sort or another. And enough organic and regenerative processes exist to show that production capacity is possible. What regenerative agriculture can't do, though, and which is an interesting and critical part of any transition, is that it can't compete against subsidized intensive agriculture. If the government pays a subsidy to generate a product, the market is not working. The government has stepped in and altered the market mechanism. If the market was pure, then what would happen is that those businesses would go out of business. Governments are frightened of even small failures in intensive agriculture, not just because they are accosted at every turn by lobbyists with deep pockets but also because they know what happens if food becomes scarce. Any disruption to society caused by food shortages is worse than an energy shortage. If there is no food on the shelves, and half your population has semi-automatic weapons, then you have a serious crisis on your hands.

A shift from intensive businesses subsidised by energy and market manipulations towards regenerative agriculture means transitioning the big companies that produce all the pesticides and fertilisers. Equally, the farmers who have been growing food intensively for a long time and are seeing their soils disappear will not want to lose money or go out of business, and they certainly don't want to be shown up as poor-quality custodians. A lot happens on a farm. It's gnarly work changing a tractor bearing, moving the livestock into the dip, checking the ewes for blight, filling the spreader with insecticide and fixing the fences. All sorts of skills are required. But not knowing that the basic resource you're dealing with is the soil and that it is your major tool is a huge disadvantage to any farmer. They will need support. These changes must happen so people can still hold their heads up whilst transitioning to a more biologically sensible way of producing food.

Legislators must be educated to face their fears and stand up to corporations manipulating their policies. Corporations themselves have to diversify their business models. As shareholders through our super funds, we need to channel our investments into smarter places and force fund managers to make smarter calls. Then, all the way down through the extension process, the landholders themselves need to come on board and change their educational profile to know what soil is all about.

This brings us to ourselves, the people, and our pets who eat the food. It's our job to understand the wasteful, intensive system of resource use that puts rare things at risk and calls it out. We can begin to make choices around food that come from smarter production. It doesn't mean becoming a vegetarian or a vegan. Still, it does mean knowing where your food comes from and, wherever possible, making choices about buying that food from sources and landholders building the carbon in their soils rather than eroding it away. We will need education. Reading this book is a good start, and from there, learn skills and techniques to understand where the food comes from.

We will have to pay more for our food during this transition. Eventually, regenerative agriculture across the board will bring the food prices back down again. However, in the transition, when we are still dealing with subsidized businesses promoted by corporations through the legislative process, we will end up paying for more sensibly grown materials. Fortunately, the process of regeneration can be quite fast. In a few years, degraded systems can become productive ones. Give them a decade or more; they'll be unrecognizable from their dusty past. And this can be done at scale.

"Ah, yes, but," says Eric, still short at a handful of rare species on his twitches list. "I want to see those rare things. I do."

And this is the nub of it. Like it or not, humanity has put itself in a place where needs must take priority over wants. Failure to meet needs where there are eight billion of us consigns us to ugly volatility and collapse. Nobody will care about the koala when the catastrophe strikes. And before you sing of regenerative agriculture as another silver bullet solution that I already said doesn't exist, the soil is where we start and where we finish. Collapse is inevitable without food. Sadly, humanity can survive without the elephant.

It is our choice. Our obligations to ourselves and each other. What our culture says about the decisions that we make. And whether or not we can rise from our lizard brains and make choices with our higher selves. We prefer to live in our fearful selves rather than rise above them. Only that's what it's going to take. Change things around to admit our mistakes and, with some sacrifices, move forward for our well-being and the well-being of our descendants. Save ourselves to give the rare things a chance.

The decisions we make in the next few decades will determine our fate. We could have argued that the next few decades would make a difference at any point in history. And they always have. We've taken a path of intensification and population growth, arguably of least resistance with the fewest tough choices. The current path started all smooth and downhill, but now it is rocky and fragile, with little resilience and resistance to the coming changes. It is time to recognize that every decision we've made in the past, conscious or otherwise, had a few decades to play out and then did. Decision-making happens, but do we carry on business as usual? This will likely cause soil to degrade further, destabilising the food supply chain with the ugly consequences that will follow. Or do we want to move towards regenerative agriculture, which we already know is closer to sustainability, which will pull down some CO_2 from the atmosphere and support the nutrient density of food, helping the diet by building on the ecology? Not only will it do those things, but it will give hope for long-term survival. Or we could put all our effort into saving the koala.

What do you think we should do?

Science Sources

This book is evidence-based. The ideas and analysis are founded on decades of experience with the science literature that I still follow closely. Where I have used specific data, it is referenced; however, to keep the text clean, I have not provided general references. If you need access to the primary sources, specific ***Evidence Reviews*** are available on The Mindful Sceptics website that you can access by clicking on or scanning the image below.

evidence reviews

scan or click the QR code for a deeper dive into the science sources for the Mindful Sceptic Guides

SCAN ME

About the Author

Mark is an ecologist, not a green one. He has built, run and sold companies, held academic posts at four universities, advised governments on environmental matters, has a PhD and still plays too much golf. His ecological research ended up in more than 70 peer-reviewed research papers and over 80 technical reports, and he supplied peer-review for more than a dozen international science journals. He lived in Africa for a decade, where he researched termites and millipedes, and once walked into a lion who snarled at him. Much of his career has been in Australia, where he won an Outstanding Teacher award from Macquarie University.

His consultancy practice advises commercial and public sector clients on assessing, monitoring, and evaluating environmental impacts, especially in carbon asset and liability management. His non-fiction books are about awareness of humans' role in nature and, as Paul Sorol, he chronicles and writes fiction for a change.

Link in bio

MORE FROM MINDFUL SCEPTICS

Uncover the power of curiosity, navigate challenges with critical thinking, and transform your perspective on how humanity can survive itself with more from the Mindful Sceptics

In posts and helpful content guides, help yourself to redefine scepticism as a tool for intelligent doubt, empowering you to question, evaluate and make informed decisions

Unlock the door to science, evidence and insights to understand the human predicament on a finite planet

Sign up now for exclusive access at

mindfulsceptics.info

MINDFUL SCEPTIC GUIDES

Abatement

Biodiversity loss

Escaping the Malthusian trap

Farmers

Food security

Food, people and soil

Growing enough food

Healthy and sustainable diet

Sustainability

The soil carbon debt

mindfulsceptics.info

Made in the USA
Monee, IL
04 November 2026

67385879R00125